Production of Energy-Efficient Natural Gas Hydrate

Production of Energy-Efficient Natural Gas Hydrate

Guest Editors

Tao Yu
Zhenyuan Yin
Bingbing Chen
Pengfei Wang
Ying Teng

MDPI

Basel • Beijing • Wuhan • Barcelona • Belgrade • Novi Sad • Cluj • Manchester

Guest Editors

Tao Yu
Key Laboratory of Ocean Energy Utilization and Energy Conversion of Ministry of Education
Dalian University of Technology
Dalian
China

Zhenyuan Yin
Institute for Ocean Engineering
Tsinghua Shenzhen International Graduate School
Shenzhen
China

Bingbing Chen
Key Laboratory of Ocean Energy Utilization and Energy Conservation of Ministry of Education
Dalian University of Technology
Dalian
China

Pengfei Wang
Advanced Institute for Ocean Research
South University of Science and Technology
Shenzhen
China

Ying Teng
Institute of Deep Earth Science & Green Energy
Shenzhen University
Shenzhen
China

Editorial Office
MDPI AG
Grosspeteranlage 5
4052 Basel, Switzerland

This is a reprint of the Special Issue, published open access by the journal *Processes* (ISSN 2227-9717), freely accessible at: https://www.mdpi.com/journal/processes/special_issues/B2YS3617X9.

For citation purposes, cite each article independently as indicated on the article page online and as indicated below:

Lastname, A.A.; Lastname, B.B. Article Title. *Journal Name* **Year**, *Volume Number*, Page Range.

ISBN 978-3-7258-6117-0 (Hbk)
ISBN 978-3-7258-6118-7 (PDF)
https://doi.org/10.3390/books978-3-7258-6118-7

Cover image courtesy of Bingbing Chen

Contents

About the Editors

Tao Yu

Tao Yu is an Associate Professor and doctoral supervisor at the School of Energy and Power Engineering, Dalian University of Technology. He earned his Ph.D. in Department of Ocean Technology, Policy, and Environment from the University of Tokyo, Japan, and previously served as an assistant professor at the Graduate School of Science and Technology, Hirosaki University, Japan. He has long been engaged in fundamental theory and applied technology research related to CO2-enhanced oil recovery and geological storage. To date, he has published over 50 SCI-indexed papers, including 18 as first or corresponding author, with more than 1,500 SCI citations and an H-index of 24. He has been granted 7 national invention patents and 1 software copyright. Additionally, he serves as a youth editorial board member for the journal *Petroleum Geology and Recovery Efficiency*.

Zhenyuan Yin

Zhenyuan Yin is an Associate Professor with the Institute for Ocean Engineering at Tsinghua University, Shenzhen International Graduate School (SIGS). His research is on the fundamentals of gas hydrate thermodynamics and kinetics with applications in energy recovery from natural gas hydrates, carbon capture and storage (CCS), and hydrogen storage coupled with renewable energy. To date, he has published more than 70 research articles, including 7 ESI high-cited papers. He was identified as the top 2% scientist by Elsevier-Stanford (2022-2024) and has received numerous local and international awards, including the ACS I&EC Research 2023 Class of Influential Researchers, China Ministry of Science and Technology Youth Talent Program (2022), Guangdong Pearl River Youth Talent Program (2022), and AIChE Outstanding Postgraduate Research thesis award (2020) for his significant contributions to the fundamentals of gas hydrate research.

Bingbing Chen

Bingbing Chen is an Associate Professor at the Dalian University of Technology, China. He has been selected for the Outstanding Youth Exchange Program of the China Association for Science and Technology and the Outstanding Doctoral Dissertation of Liaoning Province. He was also listed among the "World's Top 2% Scientists" in 2023 and 2024. His research interests focus on carbon capture, storage, and utilization technologies, as well as on the safe and efficient exploitation of natural gas hydrates and their environmental implications. He has authored more than 40 papers in SCI-indexed journals as first or corresponding author, including five ESI highly cited papers. Dr. Chen holds over 20 national invention patents, three international patents, and two software copyrights. He concurrently serves as a member of the ICGdR Special Committee on Marine Geological Disasters and Geological Environment, and as an Editorial Board Member, Young Editorial Board Member, and Guest Editor for several SCI journals, including *Natural Gas Industry B* (Impact Factor 6.5). Dr. Chen has led more than ten national and provincial-level research projects, including the National Natural Science Foundation of China Youth Program, the Special Funding for Postdoctoral Researchers, and the First-Class Funding for Postdoctoral General Projects. His academic achievements have been recognized with multiple honors, including the First Prize for Natural Science awarded by the Chinese Society of Engineering Thermophysics and the Gold Award at the 50th International Exhibition of Inventions in Geneva.

Pengfei Wang

Pengfei Wang is a Research Associate Professor at the Southern University of Science and Technology. He obtained his Ph.D. from Dalian University of Technology in 2019, with joint training at the University of Western Australia. His research focuses on gas hydrate technologies, including natural gas hydrate extraction, hydrogen purification/storage, and carbon capture and storage. With expertise in hydrate crystal structures, thermodynamics, and advanced imaging techniques, Dr. Wang has authored over 60 SCI publications and holds 10 patents, achieving an h-index of 22. He has secured several competitive grants, such as the National Natural Science Foundation Young Scientists Fund. He also serves as a youth editorial board member for international journals including *Gas Science and Engineering*, *Carbon Neutrality*, and *Energy Reviews*.

Ying Teng

Ying Teng is an Assistant Professor at Shenzhen University. She earned her Ph.D. from Dalian University of Technology in 2019, with joint training at the University of Melbourne. Her research focuses on CO2 geological storage, multiphase flow in porous media, and gas hydrate-related technologies. With a prolific publication record, Dr. Teng has authored over 40 papers in prestigious journals such as *Renewable and Sustainable Energy Reviews* and *International Journal of Heat and Mass Transfer*, and holds multiple invention patents. She currently leads several competitive research projects, including a National Natural Science Foundation of China Young Scientists Fund and a Guangdong Provincial Natural Science Foundation project.

Preface

The pursuit of sustainable and secure energy resources is a defining challenge of our time. Among the various alternatives, natural gas hydrate has emerged as a promising energy source for the 21st century, attracting significant scientific and industrial interest over recent decades. While field production tests from both terrestrial and marine reservoirs have demonstrated the technical feasibility of extraction, the path to commercial exploitation is paved with substantial challenges. A comprehensive and fundamental understanding of gas hydrate systems, from their physical properties to their environmental implications, is paramount.

This Reprint, entitled "Production of Energy-efficient Natural Gas Hydrate", is dedicated to presenting the latest advancements in hydrate research from the global scientific community. The scope encompasses a broad spectrum of studies, including experimental investigations, theoretical developments, and numerical simulations. We have encouraged submissions on a wide range of critical topics, such as the fundamentals of hydrate phase transformation, innovative production technologies, detailed resource characterization, environmental impact assessments, flow assurance challenges, and the application of hydrate science in carbon capture and storage.

The primary aim of this Reprint is to consolidate cutting-edge knowledge and foster dialogue that will accelerate the responsible development of natural gas hydrates. Our motivation stems from the urgent need to address the technical and environmental hurdles that currently limit large-scale production.

This Reprint is intended for researchers, engineers, policymakers, and advanced students in the fields of energy resources, chemical engineering, geology, and environmental science. We hope the findings contained within will serve as a valuable reference and inspire further innovation in this dynamic and critically important field.

Tao Yu, Zhenyuan Yin, Bingbing Chen, Pengfei Wang, and Ying Teng
Guest Editors

processes

MDPI

Editorial

Production of Energy-Efficient Natural Gas Hydrate

Tao Yu

Key Laboratory of Ocean Energy Utilization and Energy Conservation of Ministry of Education, Dalian University of Technology, Dalian 116024, China; yutao@dlut.edu.cn

1. Introduction

The global energy landscape is undergoing a profound transformation, driven by the dual challenges of meeting the rising energy demand and mitigating climate change. In this context, natural gas, as a relatively cleaner-burning fossil fuel, plays a pivotal transitional role [1,2]. However, its logistical challenges, particularly in stranded or offshore fields, have spurred interest in the innovative storage and transportation technologies. Among these, natural gas hydrate (NGH), a crystalline solid where the gas molecules are trapped within a lattice of water molecules, has emerged as a promising candidate. NGH technology offers a safe and compact method for gas storage, with 1 m^3 of hydrate capable of storing approximately 160–170 m^3 of natural gas under moderate pressure and temperature conditions [3,4]. Despite this potential, the conventional production of NGH has been hampered by a critical drawback: high energy consumption during the formation process. Techniques such as stirring and spraying often require significant energy input for refrigeration and mechanical agitation, which undermines the overall energy efficiency and economic viability of the technology [5].

This energy-intensive paradigm has shifted the research focus from mere hydrate production to the energy-efficient production of NGH. The core objective is to minimize the external energy input while maximizing the formation kinetics and the ultimate gas storage capacity. Recent advancements have explored various pathways to achieve this goal. For instance, the use of novel promoters, such as surfactants like sodium dodecyl sulfate (SDS) or amino acids like L-methionine, can drastically reduce the induction time and enhance the formation rates without a proportional increase in energy expenditure [6,7]. Furthermore, the adoption of innovative reactor designs and process intensification methods, including the spray method with the optimized atomizers and the application of magnetic fields with micro/nanoparticles, has demonstrated potential in improving heat and mass transfer efficiencies [8,9]. Another promising avenue involves coupling hydrate-based processes with waste or sustainable energy sources, such as utilizing the cold energy from the liquefied natural gas (LNG) regasification to provide the necessary cooling for hydrate formation [10].

Therefore, the pursuit of energy-efficient NGH production represents a critical step towards unlocking the commercial potential of hydrate technology for natural gas storage and transportation. This endeavor requires a multidisciplinary approach, integrating the insights from thermodynamics, kinetics, reactor engineering, and process integration. This Special Issue aims to comprehensively review and analyze the recent strategies and technological innovations dedicated to reducing the energy footprint of NGH production, assessing their mechanisms, performance, and potential for scalable application in a future sustainable energy system.

2. An Overview of the Published Articles

Contribution 1 investigated the influencing factors for the pilot-scale synthesis of NGH using a spray method, aiming to advance hydrate technology for natural gas storage and transportation [11]. Through systematic experiments, the authors examined the effects of reaction temperature, atomizer type, high-pressure pump flow, and water type (deionized vs. tap water) on hydrate formation, with a fixed surfactant (SDS) and pressure condition. Key findings indicate that the optimal synthesis conditions are obtained with a temperature of −5 °C and a pressure of 5 MPa, using a conical atomizer. Lower temperatures result in denser, harder hydrates with higher gas storage capacity (up to a gas-water ratio of 1:123) and more concentrated distribution within the reactor. The aperture and flow rate of the atomizer are critical: the smaller apertures improve atomization and formation speed but risk nozzle blockage, whereas higher flow rates support continuous production. The pressure difference between the pump and the reactor also enhances the atomization efficiency. Interestingly, both deionized and tap water produce hydrates with similar gas storage capacities, suggesting the economic viability of tap water for large-scale applications.

Contribution 2 used numerical simulation to evaluate the effectiveness of an optimized step-wise depressurization method for enhancing NGH production from a multi-layered reservoir in the Nankai Trough, Japan [12]. A key finding is that compared to direct depressurization, the step-wise method can increase the cumulative gas production by more than 10% over 100 days. More importantly, it effectively mitigates the rapid, short-term surges in the gas and water production rates that occur with an instantaneous pressure drop, which can help reduce sand production issues. This research identifies the depressurization gradient as a more sensitive factor than the maintenance time at each pressure step. Based on a comprehensive analysis of the production characteristics and operational feasibility, this study recommends an optimal strategy of using a 1 MPa depressurization gradient with a 1-day maintenance time at each step. This study concludes that while the step-wise depressurization method is a beneficial enhancement, the achieved production rates are still far from the commercial levels. It is therefore suggested that this method should be combined with other techniques, such as thermal stimulation, for effective future commercial exploitation.

Contribution 3 investigated the use of ferromagnetic NiMnGa micro/nanoparticles to enhance the kinetics of NGH formation [13]. Experiments were conducted using a rotating magnetic field to disperse the particles in a sodium dodecyl sulfate (SDS) solution. The results demonstrate that NiMnGa particles significantly improve hydrate formation kinetics. The most effective concentration is 3 wt%, which shortens the induction time by 98.3%, increases gas consumption by 50.5%, and increases the gas consumption rate by 351.9% compared to a pure SDS system. This enhancement is attributed to several mechanisms: the particles provide more nucleation sites, their motion under the magnetic field enhances the mass transfer at the gas–liquid interface, and their phase-change properties help absorb the hydrate formation heat. This study concludes that NiMnGa micro/nanofluids are a highly effective and novel promoter for accelerating NGH formation.

Contribution 4 employed Computational Fluid Dynamics (CFD) and Proper Orthogonal Decomposition (POD) to analyze the complex vortex structures generated by the symmetric tracer-particle injection jets within a high-pressure natural gas pipeline [14]. The key findings reveal that the interaction between the jets and the main crossflow produces distinct vortex structures, including trailing upper/lower vortices and a large-scale re-jet vortex upstream of the collision point at higher jet velocities (≥30 m/s). POD analysis successfully decomposes the flow field, showing that the first few modes, which contain the majority of the energy, capture these dominant, large-scale coherent vortex structures. The higher-order modes are associated with the shedding of the smaller-scale vortices. This research concludes that the vortex structures generated by this symmetric jet configuration promote a uniform distribution of tracer particles in the pipeline. This is crucial for ensuring the measurement accuracy of Particle Image Velocimetry (PIV) flowmeters, providing theoretical support for their application in the natural gas pipeline flow measurement.

Contribution 5 employed Grand Canonical Monte Carlo (GCMC) and Molecular Dynamics (MD) simulations to investigate the adsorption and diffusion mechanisms of CH_4, CO_2, and N_2 in the nanopores of the Longmaxi shale kerogen [15]. The key findings reveal that gas adsorption follows the Langmuir model, with the capacity increasing with the pressure but decreasing with the temperature. The adsorption affinity follows the order of $CO_2 > CH_4 > N_2$, which is a result attributed to the stronger interactions of CO_2 with the kerogen matrix, particularly with oxygen- and nitrogen-containing functional groups. Conversely, the self-diffusion coefficient of the gases shows the opposite trend: $N_2 > CH_4 > CO_2$. The diffusion capacity decreases with increasing pressure but is enhanced by higher temperatures. For the gas mixtures, this study demonstrates competitive adsorption, where an increasing mole fraction of CO_2 or N_2 reduces the adsorption of CH_4. This insight is crucial for the strategies like CO_2-enhanced gas recovery (CO_2-EGR), as injecting CO_2 can displace the adsorbed CH_4, thereby improving the shale gas extraction efficiency. In summary, this research provides a fundamental molecular-level understanding of gas storage and transportation in the kerogen, which is vital for optimizing shale gas exploration and recovery.

Contribution 6 systematically investigated the effects of a thermodynamic promoter, Tetrabutylammonium Bromide (TBAB), both alone and in combination with various kinetic promoters (SDS, nano-Al_2O_3, L-methionine, and L-leucine), on the formation characteristics of CO_2 hydrates [16]. The results identify a 4 MPa pressure and a 5 wt% TBAB concentration as the optimal thermodynamic conditions, providing the best balance of a high CO_2 separation coefficient and significant gas consumption. Furthermore, this research demonstrates that while the single kinetic promoters are ineffective on their own, combining them with TBAB significantly shortens the induction time and increases gas consumption, with the 5 wt% TBAB + 0.1 wt% L-methionine combination proving the most effective. This study concludes that using the hybrid thermo-kinetic promoters is a promising strategy for improving CO_2 hydrate formation kinetics and separation efficiency, providing valuable theoretical support for the application of this capture technology.

Contribution 7 synthesized the key sedimentary features and controlling factors governing the gas hydrate distribution in the Shenhu area [17]. This study integrated high-resolution sequence stratigraphy with drilling and logging data to reveal that Bottom Simulating Reflectors (BSRs), as the key indicators of the presence of the gas hydrate, are predominantly distributed along the continental slope ridges and within the side-slope slip blocks, typically forming thin layers (10–40 m) at the base of the gas hydrate stability zone. The distribution of BSRs and hydrates is significantly influenced by geological features such as gas chimneys, which facilitate vertical methane migration and dynamic processes including deepwater channel migration, canyon erosion, and submarine landslides. Sedimentologically, the sand content in the area is generally low (<10%), with BSRs accumulating preferentially in the zones where the sand factor is relatively higher (4–10%). Hydrate saturation shows a positive correlation with sediment porosity, and the lithology of the hydrate-bearing strata is primarily composed of fine-grained clayey silt and silty sand. The early Pleistocene erosion and resedimentation events are identified as a critical factor leading to the heterogeneous distribution of hydrates. This research establishes a sedimentary model highlighting the slumping canyon erosion channel–deepwater fan system and emphasizes that integrating sedimentary facies analysis with structural features is essential for predicting the gas hydrate distribution, thereby providing valuable insights for future hydrate exploration and resource assessment.

Contribution 8 systematically examined the major geological risks associated with the commercial extraction of submarine NGH and proposes the corresponding technical and policy countermeasures [18]. The potential hazards are categorized into two main types: natural geological hazards, including stratum subsidence, seafloor landslides, and the greenhouse effect, and secondary geological accidents, such as sand piping, well blowout, and wellbore instability. These risks primarily arise from the dissociation of NGH during mining, which alters the physical and mechanical properties of the sediment, leading to reduced shear strength, increased pore pressure, and potential large-scale deformation or failure. To mitigate these hazards, this study proposes specific technical precautions, including optimizing the CO_2-EGR method to alleviate stratum subsidence, filling seafloor strata with foamed cement to control seafloor landslides, improving drilling technology and using replacement methods to mitigate the greenhouse effect, installing solid–liquid separators to reduce sand piping, adopting new inhibitors to lower well blockage, and using low-temperature drilling fluids to maintain wellbore stability. Furthermore, this study emphasizes the importance of policy support, suggesting the establishment of a risk identification mechanism through marine science and technology research, an international risk warning mechanism, and improved safety regulations through legal supervision. This comprehensive review underscores that while NGH represents a promising future clean energy source, its commercial exploitation requires careful risk management through integrated technological innovations and robust policy frameworks to ensure safety and environmental protection.

3. Summary and Conclusions

Based on the eight contributions presented, a comprehensive picture of the current research in NGH emerges, spanning from targeted production enhancement and innovative storage techniques to fundamental molecular understanding and critical risk assessment. The collective findings underscore a multi-faceted approach to overcoming the technical and economic challenges associated with this promising energy resource.

A significant research endeavor is dedicated to optimizing the hydrate life cycle. For production from submarine reservoirs, numerical simulations demonstrate that a step-wise depressurization strategy can boost the gas yield by over 10%, and crucially mitigate sand production by preventing rapid pressure transients, although its commercial viability would still require a combination with other methods like thermal stimulation. Conversely, for natural gas storage and transportation applications, pilot-scale studies identify the optimal spray synthesis parameters—such as a temperature of -5 °C, specific atomizer types, and the pragmatic use of tap water—to achieve a high gas storage density efficiently.

Parallel to these process optimizations, the advanced materials and promoters show immense potential for enhancing the formation kinetics. The use of ferromagnetic NiMnGa particles under a magnetic field dramatically accelerates hydrate formation by improving mass transfer and heat dissipation, while hybrid promoter systems combining thermodynamic (e.g., TBAB) and kinetic agents (e.g., L-methionine) prove highly effective for CO_2 hydrate formation, relevant for separation technologies. Underpinning these applied research efforts are the sophisticated modeling and characterization studies that have been carried out. At the molecular level, the simulations reveal the competitive adsorption and diffusion behaviors of the gas mixtures in the shale kerogen, providing a mechanistic basis for strategies like CO_2-EGR. At the equipment scale, the CFD analysis elucidates how the vortex structures in the pipelines ensure uniform particle distribution, which is critical for the accurate flow measurement—a key enabling technology for the entire gas industry.

Finally, the path to commercialization is rigorously addressed through geological and risk analyses. The research in the Shenhu area synthesizes a predictive sedimentary model, linking hydrate distribution to specific geological features like gas chimneys and sediment porosity, which is vital for effective resource exploration. Acknowledging the substantial hurdles, a systematic risk assessment categorizes the geological hazards of submarine extraction, such as seafloor instability and wellbore failure, and proposes an integrated mitigation framework encompassing technical solutions like optimized CO_2-EGR and foamed cement, alongside essential policy measures for international warning systems and safety regulations.

In conclusion, these contributions collectively highlight that the future of NGH as a viable energy source hinges on an integrated strategy. This strategy must synergize a continuous innovation in gas production and storage technologies, a deep fundamental understanding of gas behavior across scales, and a proactive, comprehensive management of the associated geological and environmental risks.

Conflicts of Interest: The author declares no conflicts of interest.

References

1. Bistline, J.E.T.; Young, D.T. The Role of Natural Gas in Reaching Net-Zero Emissions in the Electric Sector. *Nat. Commun.* **2022**, *13*, 4743. [CrossRef] [PubMed]
2. Flavin, C.; Kitasei, S. The Role of Natural Gas in a Low-Carbon Energy Economy. *Worldwatch Inst.* **2010**, *37*, 33–60.
3. Sloan, E.D. Fundamental Principles and Applications of Natural Gas Hydrates. *Nature* **2003**, *426*, 353–359. [CrossRef] [PubMed]
4. Makogon, Y.F. Natural gas hydrates—A promising source of energy. *J. Nat. Gas Sci. Eng.* **2010**, *2*, 49–59. [CrossRef]
5. Song, Y.; Yang, L.; Zhao, J.; Liu, W.; Yang, M.; Li, Y.; Liu, Y.; Li, Q. The Status of Natural Gas Hydrate Research in China: A Review. *Renew. Sustain. Energy Rev.* **2014**, *31*, 778–791. [CrossRef]
6. Zhong, Y.; Rogers, R.E. Surfactant Effects on Gas Hydrate Formation. *Chem. Eng. Sci.* **2000**, *55*, 4175–4187. [CrossRef]
7. Sadeh, E.; Farhadian, A.; Maddah, M.; Semenov, M.E.; Son, E.R.; Heydari, A.; Mirzakimov, U.Z.; Valiullin, L.R.; Varfolomeev, M.A. High Storage Capacity and Rapid Methane Hydrate Formation Using Low Concentrations of A New Surfactant: A Mimic of SDS and Amino Acid Scaffold. *Appl. Energy* **2025**, *379*, 124924. [CrossRef]
8. Kibkalo, A.A.; Pandey, G.; Pletneva, K.A.; Molokitina, N.S.; Kumar, A.; Zhingel, P. Enhanced Methane Hydrate Formation Kinetics in Frozen Particles of Biopolymer Solutions: Applicable to Methane Storage. *Energy Fuels* **2023**, *37*, 13928–13936. [CrossRef]
9. Liu, N.; Liu, D.; Xie, Y.; Hu, H. Experimental Investigation of Gas Hydrates Production in a Spraying Reactor. In Proceedings of the International Conference on Power Engineering (ICOPE-2007), Hangzhou, China, 23–26 October 2007.
10. Chong, Z.R.; He, T.; Badu, P.; Zheng, J.N.; Linga, P. Economic Evaluation of Energy Efficient Hydrate based Desalination Utilizing Cold Energy from Liquefied Natural Gas (LNG). *Desalination* **2019**, *463*, 69–80. [CrossRef]
11. Ma, Y.; Zhu, J.; Meng, Q.; Ding, C.; Teng, J.; Wang, X.; Lu, Q. Analysis of Influencing Factors in Pilot Experiment for Synthesis of Natural Gas Hydrate by Spray Method. *Processes* **2022**, *10*, 2740. [CrossRef]
12. Xue, K.; Liu, Y.; Yu, T.; Lv, J. Numerical Simulation of Optimized Step-Wise Depressurization for Enhanced Natural Gas Hydrate Production in the Nankai Trough of Japan. *Processes* **2023**, *11*, 1812. [CrossRef]
13. Zhao, Z.; Wu, Q.; Li, Z.; Meng, H.; Elhefnawey, M.; Wang, X.; Wu, Q.; Li, L.; Zhang, B. Effect of Different Concentrations of NiMnGa Micro/Nanoparticles on the Kinetics of Natural Gas Hydration. *Processes* **2023**, *11*, 3149. [CrossRef]
14. Li, L.; Lu, J.; Zhao, H.; Qiu, Y. Vortex of a Symmetric Jet Structure in a Natural Gas Pipeline via Proper Orthogonal Decomposition. *Processes* **2024**, *12*, 418. [CrossRef]
15. Xiao, S.; Liu, X.; Li, Y.; Zheng, Q.; Wang, N.; Qiao, Y.; Zhang, Y.; Yi, C. Adsorption and Diffusion Properties of Gas in Nanopores of Kerogen: Insights from Grand Canonical Monte Carlo and Molecular Dynamics Simulations. *Processes* **2024**, *12*, 1438. [CrossRef]
16. Ge, Y.; Chen, H.; Qin, R.; Yao, H.; Huang, T.; Lv, X.; Liang, H.; Shen, S. Enhanced Study of CO_2 Hydrate Formation in Marine Oil–Gas Based on Additive Effect. *Processes* **2024**, *12*, 2315. [CrossRef]
17. Huang, Y.; Yu, X.; Fu, C. Gas Hydrate Distribution Influenced by Sea Floor Slide and Canyon Erosion in the Shenhu Slope, Northern South China Sea Margin. *Processes* **2025**, *13*, 193. [CrossRef]
18. Ye, Z.; Hu, W.; Yan, Q. A Review of Potential Geological Hazards and Precautions in the Mining of Submarine Natural Gas Hydrate. *Processes* **2025**, *13*, 1669. [CrossRef]

processes

MDPI

Article

Analysis of Influencing Factors in Pilot Experiment for Synthesis of Natural Gas Hydrate by Spray Method

Yun Ma [1], Jinzhao Zhu [2], Qingguo Meng [3,4,*], Chunxiao Ding [2,5], Jinbing Teng [2], Xin Wang [1] and Qian Lu [1]

1 School of Civil Engineering, Yangtze Normal University, Chongqing 408100, China
2 Linyi Special Equipment Inspection Institute, Linyi 276002, China
3 The Key Laboratory of Natural Gas Hydrate, Ministry of Natural Resources, Qingdao Institute of Marine Geology, Qingdao 266237, China
4 Laboratory for Marine Mineral Resources, Laoshan Laboratory, Qingdao 266237, China
5 School of Marine Earth Sciences, Ocean University of China, Qingdao 266100, China
* Correspondence: mengqimg@126.com

Abstract: In recent years, the technology of storing and transporting natural gas in the form of hydrate has received a lot of attention. At present, the research on the synthesis of natural gas hydrate for the purpose of storage and transportation is still in the laboratory stage, and its synthesis process is in the design and conception stage. The influencing factors of natural gas hydrate synthesis under pilot-scale conditions are more complex. Moreover, pilot experiments are oriented to actual production, and its economic feasibility and operational convenience have higher requirements. This paper aimed to study the influencing factors of gas hydrate synthesis by spray method under pilot-scale conditions. Under specific conditions of surfactant and pressure, we carried out research on the effects of reaction temperature, different forms of atomizers, high-pressure pump flow, experimental water, and other factors. Experiments show that the optimal synthesis conditions were a temperature of −5 °C, a pressure of 5 MPa, a conical nozzle, a generated gas hydrate as the hydrate of type I structure, and a gas storage capacity of 1:123 (gas–water ratio).

Keywords: natural gas hydrate; artificial synthesis; pilot-scale experiment; spray method

1. Introduction

Natural gas hydrate is a clathrate crystalline compound formed by one or more hydrocarbon gases reacting with water at a certain temperature and pressure [1]. It is globally recognized as a new type of clean energy due to its large global reserves, wide distribution, and clean combustion. With the deepening of research, more and more scholars have found that conventional natural gas can be prepared into solid hydrate for storage and transportation. It is believed that the natural gas hydrate storage and transportation technology may replace the liquefied natural gas technology and become one of the large-scale natural gas storage and transportation methods in the future [2–6].

Since Dr. Gudmundsson et al. advocated the idea of ocean transportation of natural gas by means of natural gas hydrate (NGH) in 1996 [7], technical research on hydrate formation storage and transportation has been carried out abroad and gradually used for commercial development [8–14], such as in the case of the Marathon Oil Corporation in the United States and in certain cases in Southeast Asia with the Mitsui Engineering and Shipbuilding Co.(MES) in cooperation with six Japanese leading companies related to natural gas businesses.

However, hydrate storage and transportation technology is still in the laboratory development stage in China, and has not yet formed a large-scale commercial application [15–18]. The economical and efficient synthesis of solid hydrates is the key to the industrialization of this technology. In terms of the synthesis of hydrates, scholars at home and abroad have completed many useful explorations. Scholars have focused on two aspects of hydrate formation conditions and synthesis rates, mainly through three types of methods: strengthening test equipment, strengthening physical conditions (nanofluid, external field action), and strengthening chemical reagents to improve hydrate phase equilibrium conditions, increase gas–liquid contact area, enhance heat transfer efficiency, and improve gas solubility [19–28]. Rogers et al. first used an ultrasonic sprayer to spray water into a low-temperature pressurized reactor, reducing the diameter of the water mist droplets to about 90 μm, which enlarged the contact area between water and gas by about 60 times [29]. Xie et al. found that the hydrate synthesis efficiency was higher in continuous intakes than in intermittent intakes through experiments [30]. Zhou et al. believed that the high airflow rate and prolonged ventilation time could shorten the induction period of hydrate synthesis by about two thirds [31]. Hu et al. pointed out that the selection of nozzles was very important in the experiment of preparing hydrate by spraying and believed that a suitable nozzle could not only increase the contact area of gas and water, but also strengthen heat and mass transfer [32]. Yang et al. used the spray method to improve the binding rate of water molecules and gas molecular groups and expand the contact area to enhance the rate of hydrate formation [33]. Liu et al. inted out that spraying measures could effectively shorten the induction time of hydrate formation. When the initial water temperature was constant, the higher the initial pressure of the system, the shorter the induction time; when the initial pressure was constant, the lower the initial water temperature was, the shorter the induction time was [34]. Zhang et al. proposed a spray-enhanced continuous hydrate preparation device, which used an induction reactor and a main reactor to separate and control the induction reaction and the rapid crystallization reaction, which facilitated the efficient preparation of hydrates [35]. Hu et al. also pointed out that adding an appropriate concentration of SDS (sodium dodecyl sulfate water solution) in the spray system could raise the average hydrate formation rate 1.5 times, while enhancing the gas storage capacity of hydrate [36]. Song believed that under the conditions of temperature (275.15 K), pressure (5.0 MPa), and 500 r/min stirring driving conditions, the SH-type natural gas hydrate could be rapidly synthesized by using methylcyclohexane and methyl tert-butyl ether with the addition of 0.25% surfactant SDS or rhamnolipid [37]. Shi [38] and Lin et al. [39] studied the effect of amino acids and nanosphere promoters on hydrate synthesis and proposed that amino acids could reduce hydrate growth conditions and enhance later growth, and spiral stirring could ensure the excellent hydrate formation in the early reaction. Zhong et al. found that the critical micellar concentration (CMC) of sodium dodecyl sulfate water solution was 242 ppm at hydrate forming conditions, and above this value, the hydrate synthesis speed increased at a high speed [40].

To summarize, the synthesis of natural gas hydrate in the current hydrate storage and transportation method is still in the experimental stage, and there are very few pilot-scale experimental studies. The commercial application of this technology is not yet imminent. Based on previous research results, the SDS-based spray method to synthesize hydrate is the preferred way to promote the industrialization of gas hydrate synthesis technology. In this experiment, the experiment of artificial synthesis of natural gas hydrate by spray method in pilot scale was further analyzed, and the feasibility of the synthesis method was verified. Under the conditions of surfactant SDS and certain pressures, we studied the effects of reaction temperature, different forms of atomizers, high-pressure pump flows, experimental water, and other factors. This provides a reference for the industrialization of hydrate storage and transportation methods at a later stage.

2. Materials and Methods

2.1. Experimental Setup

As shown in Figures 1 and 2, the inner volume of the reactor was 0.248 m^3, the design pressure was 13.20 MPa, and the design temperature was −20~50 °C. The heat exchange area of the heat exchange tube in the reaction kettle was 1.47 m^2 and the volume was 0.003 m^3. The opening of the heat exchange tube was connected to the refrigerant at the bottom of the lower head of the reaction kettle. The outer body of the reaction kettle was covered with a thermal insulation layer. The pressure gauge and thermometer were installed in the middle of the side wall of the reactor. The spray inlet was about 40 mm from the edge of the top of the head on the reactor.

Figure 1. Reactor equipment.

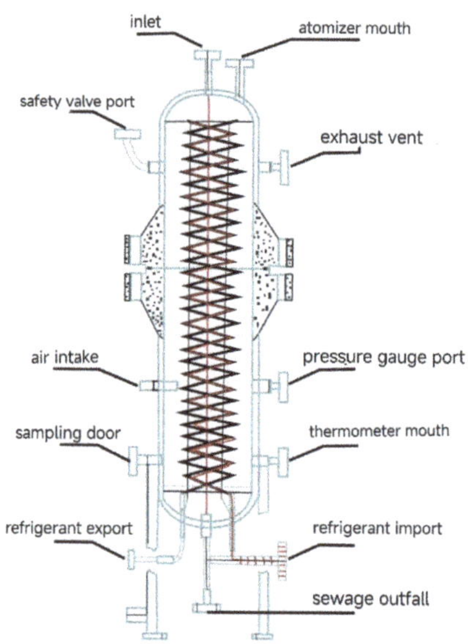

Figure 2. Schematic diagram of main structure of the reactor.

In order to verify and efficiently prepare hydrate during the experiment, four kinds of atomizers were used as a comparison, and the parameters are shown in Table 1.

Table 1. Parameter table of different atomizers.

Atomizer Types	Orifice Diameter (mm)	Flow Rate Under 0~0.2 MPa Pressure (L/h)	Jet Angle (°)	Temperature Reflex (°C)
Fine Atomizing Rotary Vane Atomizer	0.7	0~12	40	4
Stainless-steel Atomizer 1	0.79	0~35	52~58	−5
Stainless-steel Atomizer 2	2	0~100	77~82	−5
Conical Atomizer	1.5	0~114	52~65	−5

The circulating refrigeration temperature control system was a MLZ021 unit, and the refrigerant was R404A.

Two kinds of high-pressure pumps were used in this test and the parameters were in the Table 2:

Table 2. The parameter tables of the two kinds of high-pressure pumps.

Pump Types	Flow (L/h)	Greatest Pressure (MPa)
Advection pump	1.8	40
Reciprocating Piston Pump	22	60

An inVia Laser Raman Spectrometer (Renishaw, London, UK) was used from the Qingdao Institute of Marine Geology which was equipped with a low temperature cold table to ensure that there was no decomposition during the sample experimenting process. The excitation wavelength was Ar^+ laser 514.5 nm and the power was 20 mW. This spectrometer was equipped with a Leica high-performance microscope, whose confocal effect could achieve a spatial resolution of less than 1 μm laterally and about 2 μm in depth. The number of grating lines was 2400 lines/mm. The laser entered the microscope through a high-efficiency optical fiber, and a 20× objective lens was used [4].

The gas storage capacity measurement device had a vacuum degree of the vacuum pump of ≤ -95.0 KPa [41], as shown in Figure 3. The pressure sensor display range was −100 to 300 KPa and the temperature accuracy was ±0.1 °C.

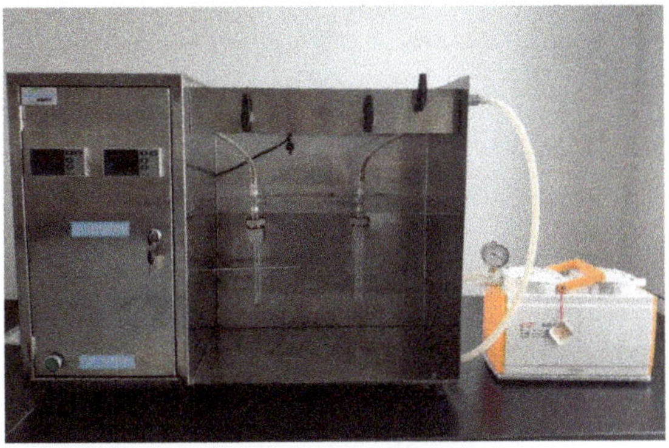

Figure 3. Gas storage capacity measurement device.

2.2. Experimental Reagents and Materials

Four kinds of reagents and materials were used in this test, they were showed in the Table 3:

Table 3. Experiment Reagents and Materials List.

Reagents and Materials	Element
Natural gas	CH_4 93.255%, $C_2H_6$3.2028%, $C_3H_8$0.6897%, Other alkane gases0.8955%, $N_2$0.8236%, $CO_2$1.1334%
Surfactant	$C_{12}H_{25}SO_4Na$ (SDS)
Deionized water	H_2O
Tap water	H_2O with some impurities

3. Synthesis Process and Experimental Observation Results

3.1. Synthesis Process

(1) First, we turned on the circulating refrigeration temperature control system, and set the temperature to the initial reaction temperature (see Table 4 for the initial temperatures under various working conditions).

Table 4. Experimental conditions and parameters.

Working Conditions		Condition 1	Condition 2			Condition 3		Condition 4			Condition 5	Condition 6
Types of Pump		Advection Pump	Piston Pump			Piston Pump		Piston Pump			Piston Pump	Piston Pump
Solution category		Deionized water	Deionized water			0.35 g/L SDS Deionized water solution		0.35 g/L SDS Deionized water solution			0.35 g/L SDS Deionized water solution	0.5 g/L SDS Tap water solution
Initial pressure (MPa)		6	6			5		5			5	5
The initial temperature (°C)		2	2			4		0			−5	−5
Ambient temperature (°C)		−6–1	−1–11			3–9		16–19			18–20	18–20
Water intake time (H)		28	2			2		2			2	2
Water intake (L)		50.4	40			40		40			40	40
Test times		3	2 times per nebulizer			3		3			3	3
Sample detection method		Raman spectroscopy	Raman spectroscopy			Raman spectroscopy		Raman spectroscopy			Raman spectroscopy	Raman spectroscopy
Atomizer		Conical	Conical Atomizer	Stainless-steel Atomizer 1	Stainless-steel Atomizer 2	Fine Atomizing Rotary Vane	Conical Atomizer	Stainless-steel Atomizer 1	Stainless-steel Atomizer 2	Conical Atomizer	Conical Atomizer	Conical Atomizer
Observation results	Hydrate formation and combustion	No	No			Yes, small flame and short burning time	Yes, small flame and burns instantaneously	Little, no flame		Yes, small flame and short burning time	Yes, big flame and lasts a long time	Yes (type I structure) big flame and lasts a long time
	State	Whole ice cubes	Colorless or pale yellow, translucent ice cubes and ice residues			Soft ice-like solid	Snow-like, slightly watery	Loose snow		Soft ice-like solid	Hard solid ice	Hard solid ice
	Distribution	At the bottom of the reactor	At the bottom of the reactor			Inside the atomizer, the water spray gap and around the atomizer, blocking the atomizer	Permeates the entire space in the reactor	At the bottom of the reactor		Extends from one side of the atomizer to the other side of the barrel	From the cylinder wall on one side of the atomizer to the spray range and the bottom of the head of the reactor	From the cylinder wall on one side of the atomizer to the spray range and the bottom of the head of the reactor

(2) After the pressure was reduced to 6–7 MPa by the pressure reducer of the cylinder group, the natural gas was slowly filled into the reaction kettle until the pressure

reached the initial reaction pressure (see Table 4 for the initial pressure under various working conditions).

(3) During the process of slowly filling the reaction kettle with natural gas, the discharge valve was opened slightly. On the one hand, it replaced the air in the reactor, and on the other hand, it helped to increase the pressure of the reactor. It was slowly filled for about 1 min then the discharge valve was closed. The air in the reactor was exhausted at this time.

(4) The change of the pressure gauge of the reaction kettle was observed. When the pressure of the pressure gauge reached the initial pressure required for the experiment, we stopped feeding the natural gas. After 30 min of stabilization, we adjusted the intake air so that the pressure value indicated by the pointer of the pressure gauge was the pressure required for the experiment.

(5) The high pressure pump was turned on, the solution (deionized water, or SDS deionized water solution or SDS tap water solution) pumped into the atomizer (a valve was installed at the inlet end of the atomizer) and then entered the reactor.

(6) The thermometer of the reaction kettle was observed constantly and the circulating refrigeration temperature control system was kept running automatically.

(7) In the process of entering the solution, the changes of temperature and pressure in the reactor were observed carefully.

(8) The temperature of the reaction kettle was kept at the temperature of step (1), after the completion of water feeding, it was left to stand for 1 h. When the pressure in the reactor did not change within 30 min, the reaction was terminated.

We observed, recorded, and measured the results, and each group of experiments was repeated three times.

3.2. Product Detection Method—Raman Spectrometry

The samples were placed in a cold stage at −190 °C. The laser excitation wavelength was 514.5 nm, the exposure time was 10 s, the scanning range was 100–4000 cm^{-1}, the slit width was 65 µm, and the 20× telephoto objective was used. Taking the sample detection of working condition 6 as an example, the Raman spectrum of the synthesized natural gas hydrate after measurement is shown in Figure 4.

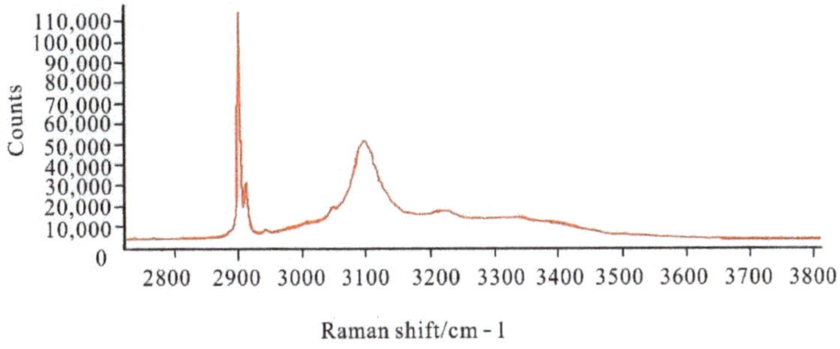

Figure 4. Laser Raman spectrum of gas hydrate.

The stretching vibration peaks of C-H bond (methane) and O-H bond (bound water) in the sample were around 2910 cm^{-1} and 3105 cm^{-1}, respectively. The C-H bond split two peaks at 2901 cm^{-1} and 2912 cm^{-1}. Compared with the C-H bond Raman spectrum peak of pure methane gas at 2916 cm^{-1} [42], both peaks of the hydrate sample were shifted to the left, mainly due to the perturbation of the electrostatic field of the water molecules forming the cages. The Raman peak of methane split around 2916 cm^{-1} due to the different spatial structures of the large cage and the small cage and the attraction to methane molecules being slightly different. The peak at 2901 cm^{-1} represents the frequency of methane molecules in the large cage, 2912 cm^{-1} represents the frequency of methane molecules in the small cage. Among them, the Raman intensity (peak area) ratio of the large cage and the small cage was basically 3:1, which is in line with the theoretical spatial structure that each unit cell of the I-type structure hydrate crystal consists of 6 large cages and 2 small cages.

The spectral peak near 3105 cm^{-1} should be the stretching vibration peak of the O-H bond, which is the vibration peak of water in hydrate. In the interaction between the methane molecule and the water molecule, the spectral peak of the water molecule also shifted. Meng et al. measured the Raman peak position of water in SDS-methane hydrate at around 3088 cm^{-1}, which is slightly different from our experimental results [43].

3.3. Test Methods for Gas Storage

The gas storage capacity testing device was self-designed, and the design principle was based on the patent of Ye et al. [41]. The vacuum volume method was used to measure the gas storage capacity of hydrates. Taking the sample of working condition 6 as an example, the gas storage capacity of the obtained synthetic natural gas hydrate was 1:123 (gas–water ratio).

The volume of standard tube A was $V_0 = 30$ mL. The pressure after vacuuming before the test was $P_a = -95.3$ kPa. Valve 1 and valve 2 were closed, and the pressure after opening valve 3 was $P_b = -13.5$ kPa. The standard pipe gas pressure was $P_0 = P_a - P_b = -95.3\ \text{kPa} - (-13.5\ \text{kPa}) = -81.8$ kPa. The liquid nitrogen was removed and the pressure value after the temperature rose to room temperature was $P_c = 8.7$ kPa. The gas pressure in standard pipe A was $P_1 = P_a - P_c = -95.3\ \text{kPa} - 8.7\ \text{kPa} = -104$ kPa. The volume of natural gas released after hydrate decomposition was $V_1 = (-104) \times 30/(-81.8) = 38.14$ mL. The weight of sample tube B (sample tube + water remaining after hydrate decomposition) was $W_1 = 38.9599$ g. The weight of the sample tube B after the moisture in the sample tube was dried was $W_2 = 38.6503$ g. So we got the weight $W_{H2O} = W_1 - W_2 = 0.3096$ g. The density of water at 20 °C was $\rho_{H2O} = 0.998\ g/cm^3$. The volume of decomposed water in the hydrate was $V_{H2O} = W_{H2O}/\rho_{H2O} = 0.3102$ mL. Finally, we calculated that $S = 38.14\ \text{mL}/0.3102\ \text{mL} \approx 123$.

3.4. Working Conditions and Experimental Observations

We repeated the experimental steps of 2.1 with different pumps, different solutions, and different sprayers. The experimental conditions were set and the observed results are shown in Table 4.

Condition 1: The basic situation of this condition is shown in Table 4. It should be emphasized that the flow rate of the advection pump was too small, and the water sprayed through the atomizer could not achieve the atomization effect, and only entered the reaction kettle in the form of water droplets. After the experiment was completed, there were integral ice cubes in the kettle and no effective hydrate was detected.

Condition 2: In view of the situation of the working condition 1, the plunger pump with large flow replaced the previous one and matched the conical atomizer, stainless-steel atomizer 1, and stainless-steel atomizer 2. Other conditions remained unchanged, and each atomizer was tested twice. The samples obtained in the experiment were all colorless or pale yellow, translucent ice cubes, and ice slag. No valid gas hydrates were detected in the samples.

Condition 1 and Condition 2 showed that increasing the flow rate of the pump and matching with the atomizer could achieve a good atomization effect. But no active agent was added in both working conditions which test reagent were both the atomized deionized water and the methane. The direct synthesis of atomized deionized water and methane gas had a very slow or almost no reaction, and the effect was very poor. Therefore, in the latter cases of 3, 4, 5, and 6, the pump was improved and the surface active agent (SDS) was added. Continuing the experiment, we found that:

Condition 3: We added surfactant (SDS) and chose four different atomizers. After the experiment, hydrates were formed in the reaction kettles, which shows that the surfactant plays an important role in the laboratory synthesis of hydrates. Table 5 lists the parameters of the four types of atomizers. These parameters combined with Table 4, demonstrate that the smaller the atomization particle size, the more sufficient the contact with the gas, and the shorter the time to generate hydrate. However, when the hydrate is produced industrially, the atomized particle size should be carefully selected to prevent the hydrate form too fast to block the nozzle. Of the four nebulizer types tested, the conical nebulizer was the most suitable for the more sustained and efficient synthesis of solid hydrates.

Table 5. Comparison of atomizer parameters and experiment results.

Working conditions		Condition 3			
Types of pumps		Piston Pump			
Solution category		0.35 g/L SDS deionized water solution			
Initial pressure (Mpa)		5			
Initial temperature (°C)		4			
Atomizer types		Fine atomizing rotary vane high pressure atomizer	Conical Atomizer	Stainless-steel Atomizer 1	Stainless-steel Atomizer 2
Nebulizer pressure conditions (Mpa)		0–0.5	0–0.2	0–0.2	0–0.5
Atomizer orifice diameter (mm)		0.7	1.5	0.79	2
The maximum open diameter of the atomizer (mm)		0.5	1.0	0.64	1
Atomizer jet angle (°)		40	52–65	52–58	77–84
Atomizer flow (L/H)		0–20	0–114	0–35	0–150
Experimental results	Hydrate formation time	Instant generation, extremely fast	Fast	Slowly	
	Hydrate state	Hard	Hard	Loose snowflake	
	Whether hydrate is formed inside the atomizer	Yes	No	No	
	Does the atomizer nozzle generate hydrates	Yes	No	No	
	Is there a blockage	Yes	No	No	
	Gas storage	1:145	1:35	Low	

The difference between the experimental conditions of working condition 4 and working condition 5 was in the setting of the initial temperature of the reaction. After the experiment, it was found that the hydrate solid generated by the initial reaction temperature of −5 °C in working condition 5 was harder and the gas storage capacity was higher. This temperature was more suitable than the initial reaction temperature of 0 °C in Case 4.

The difference between the experimental conditions of condition 5 and condition 6 was in the reaction solution. Condition 5 had SDS deionized water solution and condition 6 had SDS tap water solution. The temperature was set at the optimal reaction temperature −5 °C which was verified by this experiment. After the experiment, solid hydrates were well synthesized, and the gas storage capacity was not much different. The gas storage capacity of condition 5 was 1:121, and the gas storage capacity of condition 6 was 1:123.

Figure 5 shows curves of temperature and pressure change with time in the reactor on condition 1 to condition 6.

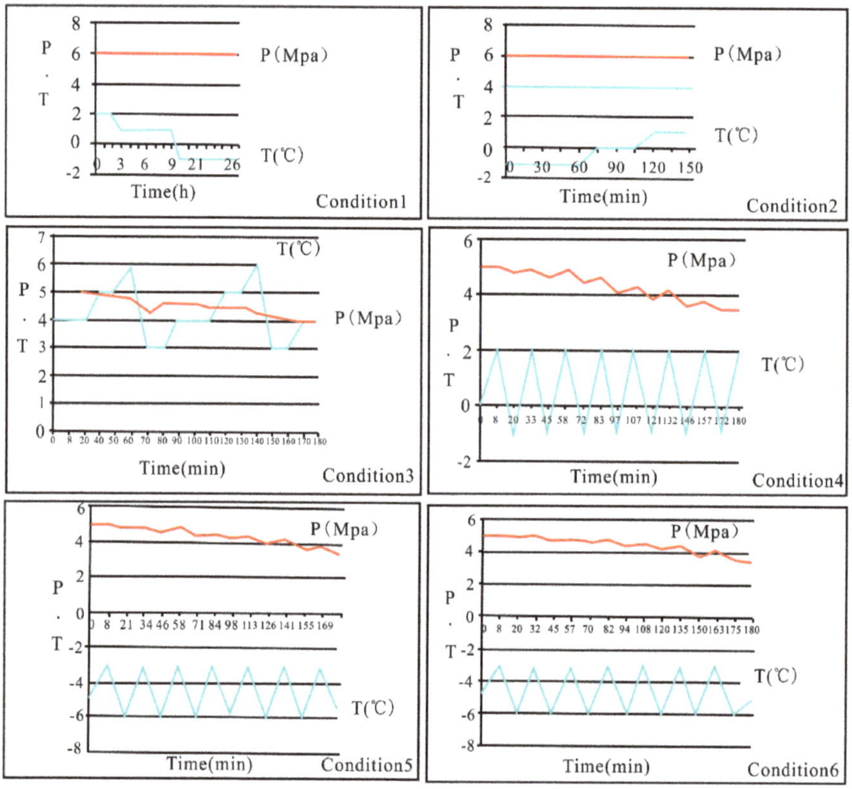

Figure 5. Curves of temperature and pressure change with time in the reactor.

Condition 1: The pressure was constant at 6 Mpa. Because the ambient temperature was reduced, the temperature in the kettle decreased from 2 °C to 1 °C and decreased to below 0 °C at 9 h. Ice began to form in the kettle, and the temperature was further reduced to –1 °C until the end of the experiment.

Condition 2: The pressure was still constant at 6 Mpa. Since the ambient temperature was lower than −1 °C, the temperature control system of circulating refrigeration did not started, so even though the initial reaction temperature was set at 2 °C, the actual reaction temperature was still not –1 °C. However, the ambient temperature increased later. The temperature in the kettle increased from −1 °C to 0 °C one hour after the experiment started, and increased to 1 °C at 105 min.

Condition 3: The temperature in kettle fluctuated within the range of 4 °C–6 °C. Since the ambient temperature was higher than the temperature in the kettle, the circulating refrigeration system was opened to ensure the set temperature. However, the intermittent opening of the circulating refrigeration system caused a certain fluctuation of the natural gas pressure in the reactor.

Condition 4: The frequency of temperature change was higher than that of condition 3 because the starting time interval of the circulating refrigeration system was shortened, and the pressure drop was slightly greater than that of condition 3. On the one hand, the temperature difference between the inside and outside of the reactor became larger due to the increase of the external ambient temperature. On the other hand, it also showed that the heat generated in the reaction process was higher than that in condition 3.

Conditions 5 and 6: The reaction temperature fluctuated within the range of −6 °C–−3 °C, and the pressure tended to decrease gradually and fluctuated with the change of temperature during the reaction process and the standing process.

Figure 6 shows the synthetic products and their distribution and state in the reactor curves of temperature and pressure changes with time in the reactor under condition 1 to condition 6.

Figure 6. The synthetic products and their distribution and state in the reactor.

4. Results and Discussion

The results from previous research and the results of this experiment both show that when the pressure is 5 Mpa, it is the ideal pressure for synthesizing hydrates. The discussion of other influencing factors in this paper was set under this pressure to ensure that the relevant unified laws were revealed to their greatest extent. When the temperature of the thermometer was 4 °C, 0 °C, and −5 °C, we carefully observed and analyzed the experimental results. When the reaction was completed, we released the pressure. We observed the pointer of the pressure gauge to see if it would return to zero. If it returned to zero, it meant that there was no internal gas to continue to discharge, and the hydrate was either not decomposed or the decomposition was completed. If the pressure gauge could not return to zero in time, the pointer vibrated near zero which indicated that not only the remaining natural gas in the inner space of the reactor was being discharged, but also that the natural gas was released by natural gas hydrate. We then closed the valve to stop the exhaust and observed the pressure gauge. If the pressure gauge started to increase slowly, it meant that the natural gas in the hydrate was still being released. When the pointer of the pressure gauge was near a certain pressure, the pressure no longer rose significantly which indicated that the decomposition of hydrate had stopped.

4.1. Influence of Temperature on the Formation, Decomposition, Distribution, and Gas Storage of Hydrate

When the pressure was 5 Mpa, the field tests of working conditions 3, 4, and 5 with the same atomizer, plunger pump, and solution were compared. Table 6 shows the reaction results under different temperature conditions.

Table 6. Comparison for field observation.

<table>
<tr><th colspan="2">Working Condition</th><th>Condition 3</th><th>Condition 4</th><th>Condition 5</th></tr>
<tr><th colspan="2">Initial Temperature (°C)</th><td>4</td><td>0</td><td>−5</td></tr>
<tr><td colspan="2">Hydrate state</td><td>Snow-like, slightly watery</td><td>Soft ice-like solid</td><td>Relatively hard solid ice</td></tr>
<tr><td colspan="2">Burning state</td><td>Instant flame</td><td>Small flames, short duration</td><td>Big flames, long lasting</td></tr>
<tr><td rowspan="4">Decompr-ession</td><td>Initial pressure relief</td><td>Pressure is around 0 Mpa, the pointer does not swing back to zero and the hydrate is still released</td><td>Pressure is around 3.3 Mpa. Slight cracking sound, the hydrate begins to decompose</td><td>Pressure is around 3 Mpa. Clear cracking sound, the hydrate begins to decompose</td></tr>
<tr><td>Close the valve</td><td>Pressure rises and hydrate continues to decompose</td><td>Stable pressure</td><td>Stable pressure</td></tr>
<tr><td>Stable pressure</td><td>Pressure = 1.0 Mpa</td><td>Pressure = 0 Mpa</td><td>Pressure = 0 Mpa</td></tr>
<tr><td>Open the lid</td><td>A large amount of gas is ejected</td><td>A moderate amount of gas is ejected</td><td>A small amount of gas is ejected</td></tr>
<tr><td colspan="2">Hydrate distribution</td><td>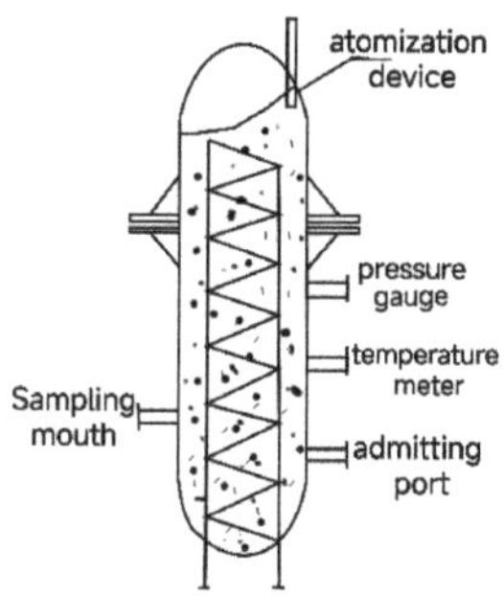

Permeates the entire space in the reactor</td><td>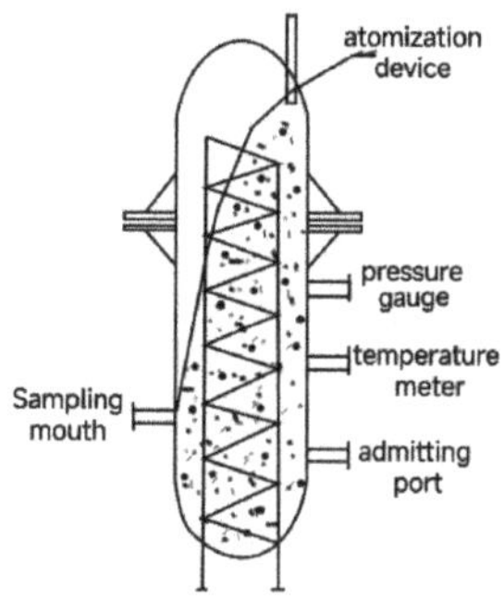

Mainly extends from one side of the atomizer to the other side, and the heat exchange tube on the sampling port side is not covered by any hydrate</td><td>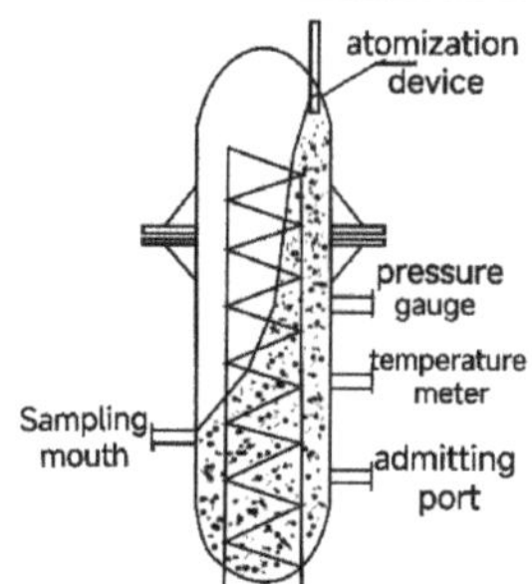

Distributed from the cylinder wall on one side of the atomizer to the spray range of the atomizer and the bottom of the head of the reactor</td></tr>
<tr><td colspan="2">gas storage</td><td>1:35</td><td>1:80</td><td>1:121</td></tr>
</table>

The observation results show that when the pressure is constant, temperature is one of the most important factors affecting the synthesis state, distribution, gas storage, and decomposition of hydrate.

(1) The compactness of the synthesized hydrate was affected: the higher the reaction temperature, the looser the generated hydrate; the lower the temperature, the denser and harder.

(2) The gas storage capacity was affected: In this experiment, the lower the reaction temperature, the higher the gas storage capacity of the hydrate sample; conversely, the higher the reaction temperature, the lower the gas storage capacity of the hydrate sample.

(3) The distribution of synthesized hydrates was affected: the lower the reaction temperature, the more concentrated the distribution of natural gas hydrates in the reactor; the higher the reaction temperature, the more diffuse the hydrates in the reactor space.

(4) The decomposition of the synthesized hydrate was affected: the lower the temperature, the less natural gas escaped from the hydrate during the exhaust pressure relief process; the higher the temperature, the more natural gas escaped.

4.2. Influence of Water Spray Flow and Atomization Effect of Atomizer on Gas Storage Capacity of Hydrate

In this experiment, four atomizers were selected to repeat the hydrate synthesis experiment. Table 6 shows the types and main parameters of the atomizer and also shows the experimental results under working condition 3.

We found that in the case of a certain pump or the pump body of this experiment:

(1) The smaller the nozzle hole diameter of the atomizer, the smaller the atomized particles, the better the atomization effect, and the faster the hydrate generation speed. However, the fine atomizing rotary vane high-pressure atomizer formed hydrate almost at the moment of spraying water, and also rapidly formed hydrate inside the atomizer, the gap where the atomizer sprays water outward, and around the atomizer. This blocked the atomizer and caused the reaction to terminate, so the atomization aperture should be suitable for continuous industrial production.

(2) The pore size of the atomizer nozzle affected the gas storage capacity of the generated hydrate. The smaller the pore size, the harder the hydrate formation texture and the higher the gas storage capacity.

(3) The larger the spray water flow rate of the atomizer, the faster the atomized solution is sprayed out of the atomizer, which did not cause the spray to generate hydrate around the nozzle and block the nozzle. The larger the water jet flow rate, the more hydrate that could be produced continuously, which is an influencing factor to ensure the continuous synthesis of hydrate.

4.3. Influence of the Pressure Difference between the High Pressure Pump and the Reactor on the Synthesis of Hydrate

Due to the existence of air pressure in the reaction kettle, the pressure difference when the pressure pump sprays water into the kettle is lower than the pressure difference when the pressure pump sprays water through the atomizer in the air, which is directly related to the matching selection of the pressure pump and the atomizer. In this experiment, the initial pressure in the reactor was 5 Mpa, the selected atomizer was a conical atomizer, and the pressure gauge of the pressure pump was in the range of 5–5.2 Mpa. While in the air, when the water mist was sprayed by the conical atomizer, the working pressure of the booster pump was between 0–1 Mpa. The spray effect formed by the same atomizer in the air was different from that formed under the experimental pressure of the reactor.

The greater the pressure difference between the high-pressure pump and the reactor, the greater the water flow rate of the atomizer, the smaller the spray angle, and the better the spray atomization effect. Therefore, choosing a pressure pump with a large effective pressure difference under high pressure was more conducive to the formation of hydrate.

4.4. Influence of Experimental Water on Gas Storage Capacity of Synthetic Hydrate

In this experiment, deionized water and ordinary tap water were selected for the experiment. Since the concentration of SDS aqueous solution was in the range of 0.3–1.1 g/L, there was no obvious difference in the synthesis effect of hydrate [16]. Therefore, in the experiment, the concentration of the SDS aqueous solution prepared with deionized water was 0.35 g/L. The concentration of the SDS aqueous solution prepared using tap water was 0.5 g/L. However, under the same conditions, the solution prepared with deionized water was basically clear and transparent after stabilization and the solution prepared with tap water was stabilized with a small amount of flocculent floating. It was speculated that the impurities in the tap water reacted with a small amount of sodium dodecyl sulfate. After the experiment was completed and the drainage was completed, the flocs still existed. The gas storage capacity of natural gas hydrate obtained in the experiment with 0.35 g/LSDS deionized aqueous solution was 1:121, and the gas storage capacity of natural gas hydrate obtained in the experiment with 0.5 g/LSDS tap water solution was 1:123.

Therefore, in the pilot scale, the experimental water had little effect on the gas storage capacity of the generated hydrate.

5. Conclusions

(1) When the pressure was suitable and constant, temperature was one of the most important factors affecting hydrate formation, decomposition, distribution, and gas storage. The lower the temperature, the harder and denser the hydrate was, the more concentrated the distribution, the higher the gas storage capacity, and the less the decomposition.

(2) The smaller the aperture of the atomizer, the better the atomization effect, the faster the formation of hydrate, and the greater the gas storage capacity. The greater the flow rate of water spray, the less likely it was to block the atomizer, and the more sustainable the process of generating hydrate. However, the smaller the pore size, the larger the flow rate of water spray that needed to be matched, otherwise the nozzle blockage was very likely to occur, resulting in the termination of the reaction. Further experiments are needed to obtain the proportional relationship between atomizer aperture and water spray flow.

(3) The greater the difference between the pressure provided by the high-pressure pump and the pressure in the reactor, the greater the water flow rate of the atomizer. The smaller the spray angle, the better the spray atomization effect and the better the quality of the hydrate produced. Therefore, a pressure pump with a large effective pressure difference was more conducive to the formation of hydrate.

(4) There was no obvious difference between deionized water and tap water for the synthesis of natural gas hydrate. The economical cost of using tap water as raw material was lower, and the process was simpler and easier to operate.

(5) In this experiment of six working conditions, the best conditions for synthesizing natural gas hydrate were: a pressure of 5 Mpa, a temperature of −5 °C, a high pressure pump reciprocating the plunger pump, model 4D-SY, a maximum working pressure of 63 Mpa, and a flow rate of 22 L/h. After the water spraying, it was left to stand for more than 60 min and the reaction was terminated after the pressure became stable. Then, a large number of hydrates with high purity could be obtained.

Factors such as the optimal volume of the reactor, the spray angle of the atomizer, and the number of atomizers under the pilot-scale conditions have not been considered in this experiment and it is expected that subsequent experiments could be improved.

Author Contributions: Conceptualization, J.Z., and J.T.; methodology, Q.M.; software, Q.L.; validation, C.D. and Q.M.; formal analysis, C.D. and Y.M.; investigation, Q.M.; resources, C.D.; data curation, C.D.; writing—original draft preparation, Y.M.; writing—review and editing, Y.M.; visualization, C.D. and X.W.; supervision, Q.M.; project administration, C.D. and Y.M.; funding acquisition, C.D. and Y.M. All authors have read and agreed to the published version of the manuscript.

Funding: This scientific research project was funded by the Quality and Technical Supervision Bureau of Shandong Province "Research on the Safety Technical Conditions of Synthetic Natural Gas Hydrate Storage and Transportation Containers" (Project No.: 2015KY19) and the Natural Science Foundation of Chongqing Municipality "Research on the Mechanism of Instability of Sandy Slopes Induced by Natural Gas Hydrate Decomposition" (Project No.: cstc2019jcyj-msxmX0815).

Institutional Review Board Statement: The study did not require ethical approval.

Informed Consent Statement: Not applicable.

Data Availability Statement: All data generated or analysed during this study are included in this published article.

Conflicts of Interest: The authors declare no conflict of interest.

References

1. Zhou, H.; Peng, X.; Ye, Y. *Natural Gas Hydrate*; Ocean Publishing House: Beijing, China, 2000.
2. Wu, C.; Sun, C.; Zhao, K.; Yang, J.; Chen, Y. Advances in natural gas storage and transportation technology for hydrates. *Oil Gas Storage Treat.* **2017**, *2*, 29–35.

3. Hu, G.; Ye, Y.; Zhang, J.; Liu, C.; Wang, J. New progress in hydrate storage and transportation technology—Record of the 6th International Hydra Conference. *Mar. Geol. Dyn.* **2008**, *24*, 17–23.
4. Shirota, H.; Aya, I.; Namie, S. Measurement of Methane Hydrate Dissociation for Application to Natural Gas Storage and Transportation. In Proceedings of the Fourth International Conference of Gas Hydrates, Yokohama, Japan, 19–23 May 2002; pp. 972–977.
5. Fan, S. *Natural Gas Hydrate Storage and Transportation Technology*; Chemical Industry Press: Beijing, China, 2005.
6. Chen, G.; Sun, C.; Ma, Q. *Gas Hydrate Science and Technology*; Chemical Industry Press: Beijing, China, 2007; pp. 1–2.
7. Gudmundsson, J.S.; Borrehaug, A. Frozen hydrate for transport of natural gas. In Proceedings of the 2nd International Conference on Gas Hydrate, Toulouse, France, 2–6 June 1996.
8. Katoh, H.; Fukazawa, K. Development of the Silo for Natural Gas Hydrate (NGH) Pellet. In Proceedings of the 6th International Conference on Gas Hydrates (ICGH 2008), Vancouver, BC, Canada, 6–10 July 2008.
9. Katoh, H.; Fukazawa, K. Development of the Silo for Natural Gas Hydrate (NGH) Pellet. In Proceedings of the 7th International Conference on Gas Hydrates (ICGH 2011), Edinburgh, UK, 17−21 July 2011; p. 571.
10. Mimachi, H.; Takeya, S.; Yoneyama, A.; Hyodo, K.; Takeda, T.; Gotoh, Y.; Murayama, T. Natural Gas Storage and Transportation within Gas Hydrate of Smaller Particle: Size Dependence of Self-Preservation Phenomenon of Natural Gas Hydrate. *Chem. Eng. Sci.* **2014**, *118*, 208–213. [CrossRef]
11. Shirota, H.; Hikida, K.; Nakajima, Y.; Ota, S.; Takaoki, T.; Iwasaki, T.; Ohgaki, K. Use of Hydrate to Natural Gas Transportation Introduction of Research Project. *Recent Adv. Mar. Sci. Technol.* **2002**, 161–167.
12. Eneota, C.; Odutola, T.O.; Ajienka, J.A. Maximizing Gas Utilization Using Gas to Hydrates Technology. *Am. J. Eng. Res. (AJER)* **2018**, *7*, 231–239.
13. Katoh, Y.; Uchida, K.; Takahashi, M.; Iwasaki, T.; Engineering, M. Experimental Research on Mixed Hydrate Pellet Production and Dissociation. In Proceedings of the 5th International Conference on Gas Hydrate, Trondheim, Norway, 13–16 June 2005.
14. Watanabe, S.; Takahashi, S.; Mizubayashi, H.; Murata, S.; Murakami, H. A Demonstration Project of NGH Land Transportation System. In Proceedings of the Manuscript for 6th International Conference on Gas Hydrate, Vancouver, BC, Canada, 6–10 July 2008.
15. Makahashi, M.; Kawamura, T.; Yamamoto, Y.; Ohnari, H.; Himuro, S.; Shakutsui, H. Effect of shrinking microbubble on gas hydrate formation. *J. Phys. Chem.* **2003**, *107*, 2171–2173. [CrossRef]
16. Javanmardi, J.; Nasrifar, K.; Najibi, S.H.; Moshfeghian, M. Economic evaluation of natural gas hydrate as an alternative for natural gas transportation. *Appl. Therm. Eng.* **2005**, *25*, 1708–1723. [CrossRef]
17. Belandria, V.; Eslamimanesh, A.; Mohammadi, A.H.; Richon, D. Study of gas hydrate formation in the carbon dioxide hydrogen water systems: Compositional analysis of the gas phase. *Ind. Eng. Chem. Res.* **2011**, *50*, 6455–6459. [CrossRef]
18. Lee, W.; Baek, S.; Kim, J.D.; Lee, J.W. Effects of salt on the crystal growth and adhesion force of clathrate hydrates. *Energy Fuels* **2015**, *29*, 4245–4254. [CrossRef]
19. Gudmundsson, J.; Andersson, V.; Levik, O.L.; Parlaktuna, M.M. Natural gas hydrates: A new gas transportation form. *J. Pet. Technol.* **1999**, *4*, 66–67.
20. Guo, Y.; Liang, H.; Guan, Y.; He, B. Effect of quaternal hinge salt on the equilibrium conditions of coalbed methane hydrates. *J. Process Eng.* **2017**, *17*, 873–878.
21. Cheng, X.; Shen, Q.; Ma, X.; Li, A.; Zhan, M. Overview of test mining of natural gas hydrates at home and abroad. *Guangdong Chem. Ind.* **2020**, *47*, 23–26.
22. Jiang, L. Experimental Study on Rapid Synthesis Technology of Natural Gas Hydrates. Master's Thesis, Southwest Petroleum University, Chengdu, China, 2018.
23. Lu, Q.; Song, Y.; Li, X. Experimental study on the hydride generation kinetics of cyclopentane-methane-aria water system in the foamer. *Chem. Prog.* **2016**, *35*, 3777–3782.
24. Wang, W. Experimental Study on Natural Gas Hydrate Synthesis under SDS System. Master's Thesis, Southwest Petroleum University, Chengdu, China, 2018.
25. Zhang, B.; Wu, Q.; Wang, Y. The mechanism of surfactants on the induction time of gas hydrate formation. *J. Jilin Univ. (Eng. Ed.)* **2007**, *37*, 239–244.
26. Ma, S.; Wu, Y.; Pan, Z.; Kang, J. Promote the research progress of natural gas hydrate generation factors. *Appl. Chem. Ind.* **2017**, *46*, 163–166.
27. Tang, X.; Sun, Z.; Chen, Z.; Liu, X.; Li, J.; Li, C.; Huang, H. Promoting tetrahydrofuran hydrate formation with copper mesh. *Cryog.* **2018**, *4*, 25–29.
28. Ganji, H.; Manteghian, M.; Zadeh, K.S.; Omidkhah, M.; Mofrad, H.R. Effect of different surfactants on methane hydrate formation rate, stability and storage capacity. *Fuel* **2007**, *86*, 434–441. [CrossRef]
29. Rogers, R.; Yevi, G.Y.; Swalm, M. Hydrate for Storage of Natural Gas. In Proceedings of the 2nd International Conference on Gas Hydrate, Toulouse, France, 2–6 June 1996; pp. 423–429.
30. Xie, Y.; Liu, D.; Fan, Y.; Xie, Y. Experimental study on the gas storage system of spray natural gas hydrate. *J. Shanghai Univ. Technol.* **2006**, *02*, 68–72.
31. Zhou, C.; Hao, W.; Feng, Z. The orifice bubble method shortens the induction period of natural gas hydrate formation. *Nat. Gas Ind.* **2005**, *25*, 27–29.
32. Hu, H.; Liu, D.; Xie, Y.; Yang, Q. Experimental study on fog-enhanced production of natural gas hydrates. *Xinjiang Oil Nat. Gas* **2006**, *2*, 68–72.

33. Yang, Q.; Liu, D.; Xie, Y.; Hu, H.; Xu, X.; Pan, Y. An experimental system for preparing natural gas hydrates by spraying. *Oil Gas Chem. Ind.* **2006**, *35*, 256–259.
34. Liu, D.; Pan, Y.; Zhou, W.; Hu, H.; Xu, X.; Yang, Q. Characteristics of the process of spraying natural gas hydrates. *J. Shanghai Univ. Technol.* **2007**, *29*, 132–136. [CrossRef]
35. Zhang, L.; Liu, D.; Zhong, D.; Li, G.; Guo, D. Process design of spray synthesis of natural gas hydrates. *Modernization* **2008**, *10*, 64–67.
36. Hu, H.; Liu, D.; Xie, Y.; Yang, Q. Effect of surfactants in the spray reactor on the preparation of natural gas hydrates. *Qilu Petrochem. Ind.* **2006**, *3*, 256–259.
37. Song, X. Experimental Study on the Synthesis of Natural Gas Hydrates Suitable for Ship Transportation. Master's Thesis, Dalian Maritime University, Dalian, China, 2021.
38. Shi, X. Research on High Efficiency Synthesis Technology of Natural Gas Hydrate Reinforced by Spiral Mixing. Master's Thesis, Qingdao University of Science and Technology, Qingdao, China, 2022.
39. Lin, Y.; Liu, L.; Sun, M.; Chen, C.; Zhang, G.; He, Y.; Wang, F. Rapid formation of methane hydrates with compact agglomeration via regulating the hydrophilic groups of nanopromoters. *AIChE J.* **2020**, *66*, e16296. [CrossRef]
40. Zhong, Y.; Rogers, R.E. Surfactant effects on gas hydrate formation. *Chem. Eng. Sci.* **2000**, *55*, 4175–4187. [CrossRef]
41. Ye, Y.; Liu, C.; Meng, Q.; Cheng, J. Experimental Device for Determining the Gas Storage Capacity of Natural Gas Hydrates. Patent No. 201010222113.8, 16 February 2011.
42. Xia, N.; Liu, C.; Ye, Y.; Meng, Q.; Lin, X.; He, X. Study on the method of microscopic laser Raman spectroscopy for the determination of natural gas hydrates. *Rock Mine Test* **2011**, *30*, 416–422.
43. Meng, Q.; Liu, C.; Ye, Y.; Chen, Q. Experimental study on gas storage characteristics of methane hydrates in different systems. *World Sci. Technol. Res. Dev.* **2011**, *33*, 25–28.

processes

MDPI

Article

Numerical Simulation of Optimized Step-Wise Depressurization for Enhanced Natural Gas Hydrate Production in the Nankai Trough of Japan

Kunpeng Xue [1], Yu Liu [1,*], Tao Yu [1,*] and Junchen Lv [2]

[1] Key Laboratory of Ocean Energy Utilization and Energy Conservation, Ministry of Education, School of Energy and Power Engineering, Dalian University of Technology, Dalian 116024, China; xkp@mail.dlut.edu.cn

[2] School of Mechanical and Power Engineering, Dalian Ocean University, Dalian 116023, China; henry-tianxia@hotmail.com

* Correspondence: liuyu@dlut.edu.cn (Y.L.); yutao@dlut.edu.cn (T.Y.)

Abstract: The utilization of natural gas hydrates as an alternative energy source has garnered significant attention due to their proven potential. Despite the successful offshore natural gas hydrate production tests, commercial exploitation has not been achieved. This study aims to enhance the understanding of gas production behavior through simulations from a single vertical well in the Nankai Trough and assess the effectiveness of the step-wise depressurization method for gas production using TOUGH + HYDRATE. The simulation results showed that the effective permeability for the water phase decreased as the hydrates were decomposed, and the invasion of the pore water from the underburden eliminated this effect. Compared with the direct depressurization method, the step-wise depressurization method significantly increased the cumulative gas production by more than 10% and mitigated the rapid generation of gas and water production during the moment of depressurization. The results also indicated that the depressurization gradient was more sensitive to the cumulative gas production than the maintenance time of depressurization. In view of the gas and water production characteristics coupled with the challenges in carrying out the step-wise depressurization method, it is suggested that a depressurization gradient of 1 MPa and a maintenance time of 1 day should be employed.

Keywords: nature gas hydrate; step-wise depressurization; gas production; Nankai Trough

1. Introduction

Natural gas hydrates (NGHs) are crystalline solid, nonstoichiometric compounds that form under a special environment of low temperature and high pressure [1–3]. They are primarily present in continental slopes, deep-sea, or permanent frozen strata [4–6]. They are considered as one of the most promising alternative energy sources due to the characteristics of high energy density, wide distribution, and combustion cleanliness [7–9].

The mechanism of the exploitation is to break the phase equilibrium of NGHs and extract gas from the NGH reservoir [10,11]. There are four main exploitation methods, including depressurization, thermal injection, chemical inhibitor injection, and carbon dioxide replacement methods [12,13]. The thermal injection method has poor energy efficiency due to the low energy transformation ratio [14,15]. Environmental concerns and potential reservoir pollution are major problems for the chemical inhibitor injection method [16,17]. The low replacement rate limits the development of the carbon dioxide replacement method [18,19]. Depressurization is generally believed to be the most promising method because of its simplicity, technical feasibility, and economic effectiveness [20–22]. Recent offshore NGH field tests conducted by Japan and China, during which the depressurization method was used, suggested the possibility of extracting gas from NGH reservoirs [23–26]. In 2013, Japan conducted the world's first offshore production test of

NGHs using a single vertical well. This test was successful in producing a cumulative volume of 119,500 m^3 of natural gas over a period of 6 days, with an average daily production rate of 2.0×10^4 m^3/d. It is worth noting that the production test lasted only 6 days, primarily due to the severe issue of sand production. In 2017, Japan conducted a second offshore production test of NGHs following the first test in 2013. It was reported that a total of around 2.0×10^5 m^3 of natural gas was produced over a period of 24 consecutive days (i.e., 8.33×10^3 m^3/d) during this recent test. Additionally, in 2017, China conducted a production test of NGHs in the Shenhu area of the South China Sea, achieving a cumulative gas production of 3.09×10^5 m^3 during 60 days with an average production rate of 5.15×10^3 m^3/d. In 2020, China's second production test of NGHs resulted in an improvement in the average gas production rate, with a cumulative gas production of 8.614×10^5 m^3 over 30 days and an average production rate of 2.87×10^4 m^3/d. However, the issue of sand production remains a significant concern and the gas production rates of the recent field tests were far from the commercial exploration level of 5.0×10^5 m^3/d (0.1 MPa and 289 K) [2]. Consequently, in order to achieve successful commercial gas production from offshore NGH deposits in the future, significant improvements in production efficiency are necessary, and the depressurization method must be optimized [27].

Numerous laboratory-scale studies have investigated the optimization of depressurization method for the development of the offshore NGH deposits, including the step-wise depressurization method. For example, Heeschen et al. [28] compared the effects of two gradient depressurization modes, 7.0-5.0-4.2 MPa and 9.0-7.0-5.0-4.2-3.0 MPa, on gas production behaviors using the gas production test results of Mallik as a basis. The results showed that the second mode with a small depressurization gradient had a higher cumulative gas yield. Yang et al. [29] used real marine sediments as a carrier to compare and study the production behaviors of natural gas by single-step and multi-step decompression methods and found that the multi-step method could alleviate the reservoir temperature drop. Zhao et al. [30] employed a multi-level pressure reduction approach in their study of hydrate exploitation and found that it could effectively mitigate the reservoir temperature decline. In addition, a small depressurization gradient of 0.5 MPa could increase the rate of hydrate decomposition by 18.92%. Overall, at the laboratory scale, the step-wise depressurization method may potentially improve the rate of hydrate decomposition and the cumulative gas production.

Due to the difficulties and exorbitant costs associated with field testing, the study of reservoir-scale phenomena is often carried out using numerical simulations. These simulations prove useful in determining the most efficient exploitation methods and monitoring changes in reservoir characteristics. Several specialized simulators, such as TOUGH + HYDRATE, STOMP-HYD, CMG-STARS, and MH21-HYDRES, have been developed for this purpose. Utilizing the field test data, numerous numerical simulations have been conducted to enhance gas productivity, including the utilization of horizontal wells, hot water injection, gas exchange, and permeability enhancement techniques of the NGH reservoirs. However, research on the step-wise depressurization method for gas production at the reservoir scale remains limited.

The main objective of this study was to investigate the effects of different step-wise depressurization methods on natural gas hydrate production in the Nankai Trough of Japan and identify the optimal depressurization strategy. To achieve this aim, a multi-layered NGH reservoir model was established, which provided a more accurate representation of the reservoir conditions compared to a single-layered model. The accuracy of the model was confirmed through the comparison with the actual field test data. Then, the gas and water production characteristics and changes in reservoir permeability over the course of one year were analyzed in this study. Ten scenarios were simulated to evaluate the effects of the key factors, such as the depressurization gradient and maintenance time after each depressurization step, on gas and water production. The cumulative gas and water production from different step-wise depressurization cases were also compared after 100 d of production. Additionally, the evolution of reservoir characteristics distribution by the step-wise depressurization method had also been presented in this study. The results of this research are expected to provide valuable guidance for future commercial NGH production initiatives.

2. Numerical Modeling

2.1. Numerical Simulator

The TOUGH + HYDRATE software developed by the Lawrence Berkeley National Laboratory is widely used for simulating gas production behaviors in NGH reservoirs [31]. This simulator has a good track record of accurate predictions, as confirmed by many scholars [4,32,33]. This simulator was also employed for predicting gas and water production in this study. Specifically, the equilibrium model was adopted, and all the simulation results were obtained based on the assumption that sand production has been fully controlled and the pore structure of the NGH reservoirs would not be deformed. The main control equations are the mass and energy balance equations as follows,

$$\frac{d}{dt}\int_{V_n} M^{\kappa} dV = \int_{\Gamma_n} F^{\kappa}\cdot n d\Gamma + \int_{V_n} q^{\kappa} dV, \tag{1}$$

where M^{κ}, F^{κ}, and q^{κ} represent the mass accumulation term, Darcy flux vector, and source or sink term, respectively; V_n, Γ_n, and κ represent the grid volume (m^3), grid surface area (m^2), and component, respectively; t represents time (s); and V and Γ are the volume (m^3) and surface area (m^2), respectively.

2.2. Model Construction and Domain Discretization

The production site AT1 in the Nankai Trough was chosen for the simulations. This site is located in the Daini Astumi Knoll of the Eastern Nankai Trough off the Pacific coast of Japan [24]. The NGH reservoir at this site comprise upper alternate layers consisting primarily of interbedded sand and silt, middle silt-dominated layers containing mostly silt with sand as the minor component, and lower sand-dominated layers composed primarily of thick sand sequences with little silt [22,24].

A multi-layered NGH reservoir model was utilized in this study, as depicted in Figure 1, which was constructed using a cylindrical coordinate system. The reservoir model was divided into three main sections, the overburden (OB), hydrate layer, and underburden (UB). The hydrate layer was further divided into three sublayers, the upper layer (HBL1), middle layer (HBL2), and lower layer (HBL3), corresponding to the three hydrate sublayers mentioned previously. The OB and UB were set to be 30 m thick, which has been shown to effectively eliminate the boundary effects [34]. The thicknesses of the HBL1, HBL2, and HBL3 were set to be 20 m, 8 m, and 32 m, respectively. The radial extent of the reservoir model was set to be 200 m. A production well with a length of 38 m and a radius of 0.1 m was used in this simulation and positioned in the HBL1 and HBL2 zones, as well as the upper part of the HBL3 zone in order to match the field test.

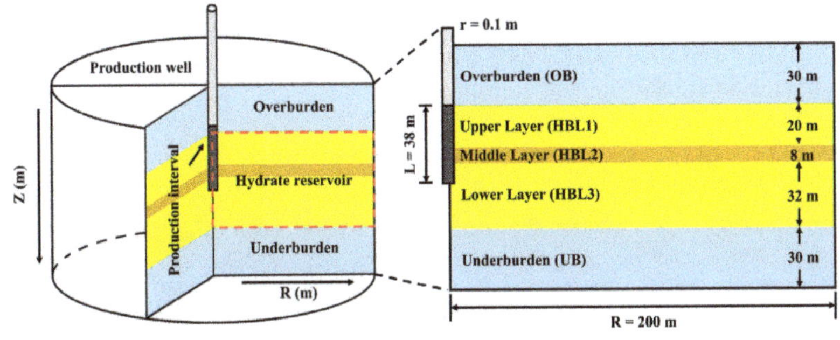

Figure 1. Schematic diagram of the established reservoir model.

In this model, a fine grid division was performed, yielding a total of 22,648 grids, with 152 grids along the radial coordinate (r) and 149 grids along the vertical coordinate (z). To ensure efficient calculation while preserving accuracy in predictions, the radial grid size was determined using logarithmic growth. It is noteworthy that the first and last grids along the radial coordinate were set to be 0.1 m to represent the wellbore from which gas and water were produced and the boundary of the model, respectively. The reaction in the hydrate layer is complex, including intricate transformations in heat conduction, hydrate decomposition, and fluid flow. To properly capture these processes, the vertical grid size for the hydrate layer was set to be 1 m, whereas the vertical grid size for the non-hydrate layer was set to be 3 m. Furthermore, to ensure an accurate representation of the reaction, the grids were meticulously fine-tuned at the interface between the layers.

2.3. Initial and Boundaries Conditions

The pressure gradient of the model was assumed to be the hydrostatic pressure gradient and calculated by the empirical formula [35],

$$P_{pw} = P_{atm} + \rho_{sw} g * (H + Z) * 10^{-6}, \quad (2)$$

where P_{pw} represents the hydrostatic pore water pressure and P_{atm} represents the standard atmospheric pressure, MPa; ρ_{sw} is the average seawater density, kg/m^3; g stands for the gravitational acceleration, m/s^2; H is the water depth, m; and Z is the depth of the sediment from the seafloor, m.

The temperature distribution was derived according to the geothermal gradient and the mean seafloor temperature as follows [22],

$$T = T_{sf} + \Delta T \cdot Z \times 10^{-3}, \quad (3)$$

where T_{sf} represents the mean temperature of the sea floor, and the value is set to be 3.5 °C according to the literature [24]; ΔT is the geothermal gradient, and the value was set to be 30.0 °C/km. In this simulation, a set of parameters based on relevant studies were used to model the reservoir, such as the porosity, initial hydrate saturation, and intrinsic permeability of each layer [10,36,37]. For example, (a) the porosity was set to be 0.40 for the whole reservoir; (b) the initial hydrate saturations of the HBL1, HBL2, and HBL3 were set to be 0.50, 0.35, and 0.60, respectively; (c) the intrinsic permeability of the reservoir was considered in relation to anisotropy, i.e., the permeabilities in the horizontal and vertical directions of the HBL1, HBL2, and HBL3 were set to be 0.30 D and 0.20 D, 0.05 D and 0.05 D, and 0.40 D and 0.30 D, respectively. The permeability of the OB and UB were set to be isotropic, with the values of 0.01 D and 1.00 D, respectively. The parameters of each layer in this model are listed in Table 1, and the reservoir parameters and conditions used in the simulations are shown in Table 2.

In order to implement the first-type boundary conditions, also known as Dirichlet boundary conditions, fixed pressure and temperature were imposed at the top and bottom of the model. The wellbore pressure was set to be 4.5 MPa, which aligned with the production pressure observed in the field. As depicted in Figure 2, after a prolonged simulation, the formation conditions remained stable, indicating that the model was successfully constructed.

Table 1. Parameters of each layer in the model.

Layer	Parameter	Value and Unit
OB	Thickness Porosity Intrinsic permeability	30 m 0.40 0.01 D
HBL1	Thickness Porosity Intrinsic permeability Initial hydrate saturation	20 m 0.40 0.30 D (horizontal), 0.20 D (vertical) 0.50
HBL2	Thickness Porosity Intrinsic permeability Initial hydrate saturation	8 m 0.40 0.05 D (horizontal), 0.05 D (vertical) 0.35
HBL3	Thickness Porosity Intrinsic permeability Initial hydrate saturation	32 m 0.40 0.40 D (horizontal), 0.30 D (vertical) 0.60
UB	Thickness Porosity Intrinsic permeability	30 m 0.40 1.00 D

Table 2. Reservoir parameters and conditions used in the simulations.

Parameter	Value and Unit
Seawater density (ρ_{sw})	1022 kg/m^3
Grain density	2650 kg/m^3
Grain specific heat	792 J/(kg·°C)
Wet thermal conductivity (sand)	2.917 W/(m·°C)
Wet thermal conductivity (silt)	1.7 W/(m·°C)
Dry thermal conductivity	1.0 W/(m·°C)
Capillary pressure model [38]	$P_{cap} = -P_0\left[(S^*)^{-\frac{1}{\lambda}} - 1\right]^{1-\lambda}$, $S^* = (S_A - S_{irA})/(S_{mxA} - S_{irA})$
P_0 [initial capillary pressure (Pa)]	10^4 Pa (sand), 10^5 Pa (silt)
λ [exponent in the capillary pressure model]	0.45 (sand), 0.15 (silt)
S_{mxA} [maximum water saturation]	1.00
Relative permeability model [39]	$k_{rA} = [(S_A - S_{irA})/(1 - S_{irA})]^n$, $k_{rG} = [(S_G - S_{irG})/(1 - S_{irA})]^{n_G}$
S_{irA} [irreducible water saturation]	0.25 (sand), 0.55 (silt)
S_{irG} [residual gas saturation]	0.01 (sand), 0.05 (silt)
n [exponent in the relative permeability model for the aqueous phase]	3.5 (sand), 5.0 (silt)
n_G [exponent in the relative permeability model for the gas phase]	0.01 (sand), 0.05 (silt)
p_{atm} [standard atmospheric pressure (MPa)]	0.101325 MPa
g [gravitational acceleration (m/s^2)]	9.8 m/s^2

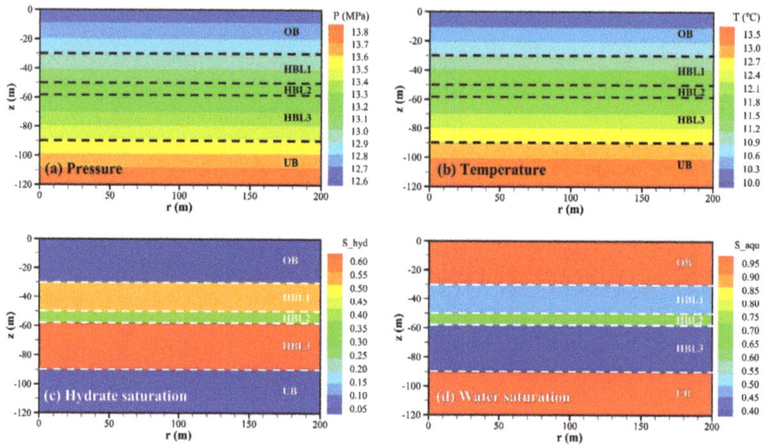

Figure 2. Initial state distribution of the reservoir parameters: (**a**) pressure distribution; (**b**) temperature distribution; (**c**) gas hydrate saturation distribution; and (**d**) water saturation distribution.

3. Direct Depressurization Method

In this section, the accuracy of the model was verified by comparing the simulation results with the field data collected during the first test of the Nankai Trough. To further investigate the characteristics of gas and water production, a year-long simulation by the direct depressurization method was conducted.

3.1. Model Validation

To ensure the accuracy of the simulations, a 6-day simulation was conducted, as illustrated in Figure 3. The gas production rate remained at approximately 2.0×10^4 m^3/d, which was consistent with the field data from the Nankai Trough with the exception of the first day. The unusually high gas production rate on the first day was caused by an instantaneous pressure drop from 13.5 MPa to 4.5 MPa, resulting in a slightly higher cumulative gas production than the actual value. Given the similar results observed in other studies [22,37,40,41], it could be concluded that the model was accurate and has been successfully established. Based on the successful establishment of the model, a one-year simulation by the direct depressurization method was conducted using this model.

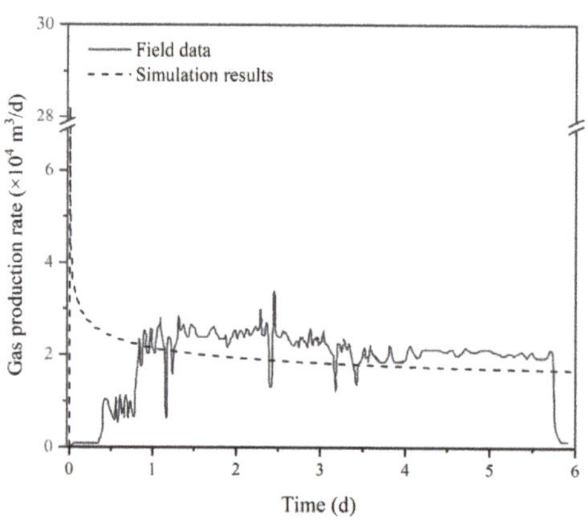

Figure 3. Comparison of the simulation results and field data.

3.2. *Characteristics of Gas and Water Production*

Figure 4 shows the evolution of gas and water production rates, as well as cumulative gas and water production, during the first year of the simulation. The results indicated that the gas production rate decreased quickly from its initial value of 2.82×10^5 m^3/d to approximately 1.6×10^4 m^3/d, which was due to the initial sudden pressure drop. The gas production rate then gradually decreased, reaching a value of 1.15×10^4 m^3/d at the end of the first year. This decreasing trend was attributed to the slow pressure propagation, which hindered the decomposition of hydrates [41]. In contrast, the water production rate remained stable at 0.6×10^4 m^3/d throughout the year. This value was significantly higher than the field test, and this phenomenon was attributed to the use of a waterproof device in the field test that was not accounted for in the simulation.

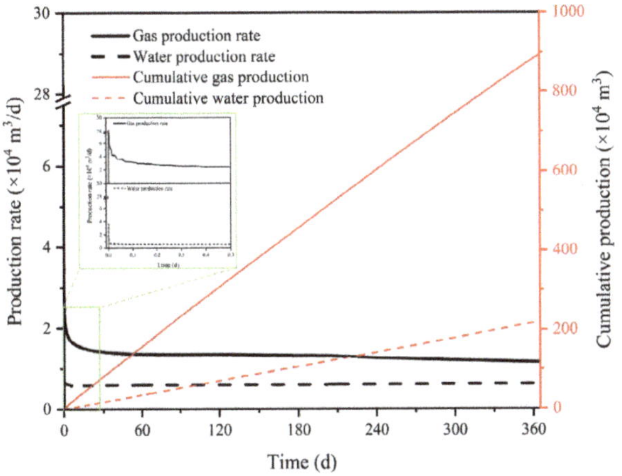

Figure 4. Gas and water production behaviors during the first year.

3.3. *Evolution of the Reservoir Permeability*

Numerous studies have shown that reservoir permeability is a critical factor controlling gas–water flow in reservoirs, which, in turn, is crucial for gas hydrate exploitation. In the TOUGH + HYDRATE simulator, the flowing phase is based on the Darcy's law and divided into aqueous phase flow and gaseous phase flow, with the effective permeability represented by the rock intrinsic permeability multiplied by the relative permeability of the aqueous/gaseous phase. In this simulation, the modified version of Stone's three-phase relative permeability method was chosen as the relative permeability model, with specific parameters shown in Table 2. The evolution of effective permeabilities in aqueous and gaseous phase flows are illustrated in Figure 5. During the exploitation process, the effective permeability of the aqueous phase was found to decrease, and the affected area gradually expanded. This phenomenon was beneficial to prevent excessive water production from the production well. The HBL2 was found to exhibit the smallest decrease in effective permeability. Additionally, it was observed that there was no significant decrease in the effective permeability of the aqueous phase in the lower part of the HBL3, hence attention should be paid to the flow of water from the UB into the production well. Regarding the effective permeability of the gaseous phase, it was observed that after gas hydrate decomposition the effective permeability increased, which facilitated the flow of gas into the production well. In addition, the effective permeability decreased as the distance from the production well increased.

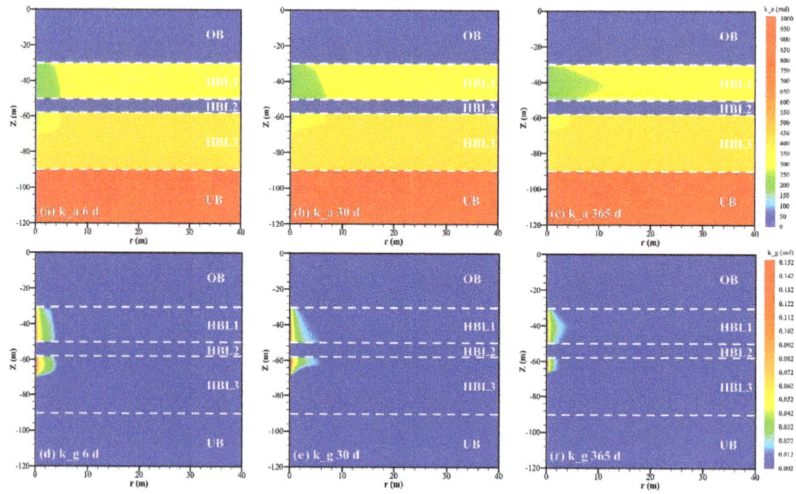

Figure 5. Evolution of the effective permeabilities in aqueous and gaseous flows during gas hydrate exploitation: (**a–c**) the effective permeability of aqueous phase at day 6, day 30, and day 365; and (**d–f**) the effective permeability of gaseous phase at day 6, day 30, and day 365.

4. Optimized Step-Wise Depressurization Method

The direct depressurization method can lead to (a) the emergence of low-temperature zones near the production well [42,43], which negatively impacts the production of nature gas, and (b) large gas and water production in a short time, which may result in sand production issues. In order to address these challenges, a step-wise depressurization method was proposed in this work. To thoroughly and clearly study the effect of the step-wise depressurization method, we have designed 10 sets of cases, each utilizing a different approach to reach a production pressure of 4.5 MPa. Specifically, Case 0 was used as a control group and the direct depressurization method was applied. Cases 1–3, 4–6, and 7–9 utilized the depressurization gradients of 3 MPa, 2 MPa, and 1 MPa, respectively, with the maintenance times of 24, 12, and 6 h after each depressurization stage. The schematic diagram of the step-wise depressurization patterns was shown in Table 3. Additionally, a summary of the different cases and their comparative analysis, aiming to study various influencing factors, was provided in Table 4. Furthermore, the evolution of the distribution of reservoir characteristics was monitored and presented visually.

Table 3. The schematic diagram of the step-wise depressurization patterns.

Case	Depressurization Step	Depressurization Process	Maintenance Time (h)	Depressurization Time (d)
Case 0 (Reference case)	1	13.5 → 4.5	None	None
Case 1	3	13.5 → 10 → 7 → 4.5	24	3
Case 2	4	13.5 → 10 → 8 → 6 → 4.5	24	4
Case 3	7	13.5 → 10 → 9 → 8 → 7 → 6 → 5 → 4.5	24	7
Case 4	3	13.5 → 10 → 7 → 4.5	12	1.5
Case 5	4	13.5 → 10 → 8 → 6 → 4.5	12	2
Case 6	7	13.5 → 10 → 9 → 8 → 7 → 6 → 5 → 4.5	12	3.5
Case 7	3	13.5 → 10 → 7 → 4.5	6	0.75
Case 8	4	13.5 → 10 → 8 → 6 → 4.5	6	1
Case 9	7	13.5 → 10 → 9 → 8 → 7 → 6 → 5 → 4.5	6	1.75

Table 4. Summary of the research cases.

Group	Cases	Objectives
Group A	Cases 0–3	
Group B	Cases 4–6	Depressurization gradients
Group C	Cases 7–9	
Group D	Cases 1, 4, 7	
Group E	Cases 2, 5, 8	Maintenance time
Group F	Cases 3, 6, 9	

4.1. Comparison of the Production Behaviors with Different Depressurization Gradients

The production characteristics of gas and water over a 30-day period for Cases 0–3 are displayed in Figure 6, where Figure 6a–d represents the gas production rate, water production rate, cumulative gas production, and cumulative water production, respectively. It can be observed from Figure 6a,b that each time a depressurization operation was carried out the gas and water production rates increased sharply. The peak gas production rate of Case 0 had reached 28.20×10^4 m^3/d, while those for Cases 1, 2, and 3 of the step-wise depressurization patterns were 10.52×10^4 m^3/d, 7.80×10^4 m^3/d, and 5.28×10^4 m^3/d, respectively. It was evident that the step-wise depressurization method reduced the sudden large surge in gas production rate from the direct depressurization method by breaking it down into multiple smaller surges. Compared to Case 0, the peak gas production rate of Case 3 decreased by 81.38%. Furthermore, when each step-wise depressurization pattern reached the final production pressure, the gas production rate was higher than that of the direct depressurization method. In terms of water production rate, Case 0 had a peak water production rate of 3.63×10^4 m^3/d, whereas those for Cases 1, 2, and 3 of the step-wise depressurization patterns were 1.61×10^4 m^3/d, 1.30×10^4 m^3/d, and 0.98×10^4 m^3/d, respectively. It can be understood that when the water production rate was high, it is easy for sand to be carried along and flow toward the production well. Therefore, it is necessary to avoid such a situation, and the step-wise depressurization method could be used to mitigate this problem to some extent. Case 3, in particular, showed a decrease of 84.02% in peak water production rate compared to Case 0. The cumulative gas and water production of Cases 0–3 are shown in Figure 6c,d. It can be observed that the cumulative gas production of the step-wise depressurization method was higher than that of the direct depressurization method, which confirmed the conclusion drawn from Figure 6a that the gas production rate of the step-wise depressurization method was higher than that of the direct depressurization method. When the gradient of the step-wise depressurization pattern was 1 MPa (Case 3), the increase in cumulative gas production became less significant compared to Case 2. As is well known, water production increases with the increase in gas production [44–46]; however, in Case 3, the cumulative water production was not as high as that in Case 2. This could be attributed to the small pressure gradient during each depressurization stage, which resulted in a weaker stimulation of water flow and a slower influx into the production well. In general, the smaller the gradient in the multi-stage depressurization pattern becomes, the more favorable it is for gas production.

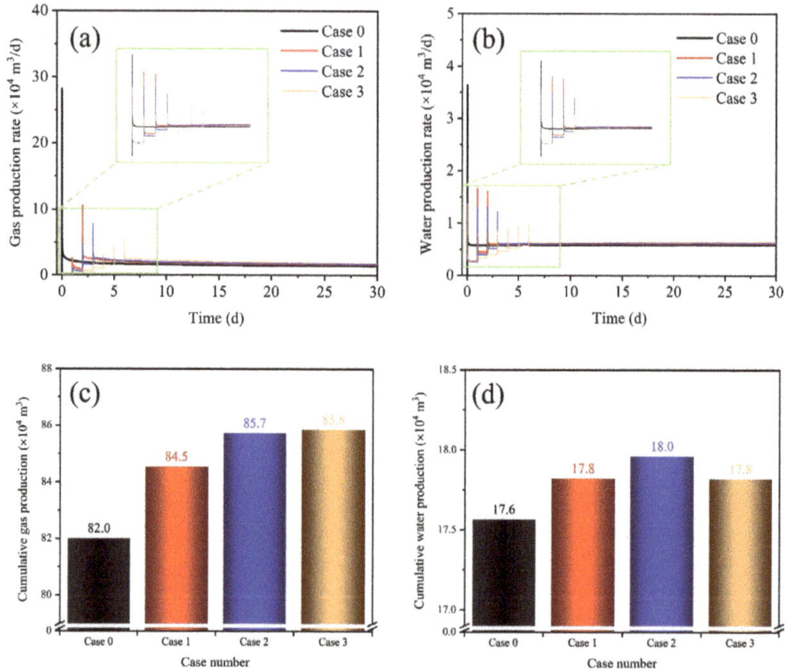

Figure 6. Gas and water production characteristics in Cases 0–3: (**a**) gas production rate; (**b**) water production rate; (**c**) cumulative gas production; and (**d**) cumulative water production.

4.2. Comparison of the Production Behaviors with Different Maintenance Times

The previous section focused on the impact of the depressurization gradient on the production behavior by the step-wise depressurization method. This section will primarily analyze the effect of maintenance time. The gas and water production behaviors are shown in Figure 7. Cases 3, 6, and 9 all feature the same depressurization gradient of 1 MPa, with different maintenance times of 24 h, 12 h, and 6 h, respectively. From Figure 7a, it can be observed that the peak gas production rate of Case 9 (6.10×10^4 m^3/d) was slightly higher than that of Case 3 (5.28×10^4 m^3/d), and the gas production rates of all the three cases eventually approach a similar value. As shown in Figure 7b, the peak water production rates were almost identical among Cases 3, 6, and 9, with the values of 0.97×10^4 m^3/d, 0.97×10^4 m^3/d, and 0.98×10^4 m^3/d, respectively. This indicated that the peak value of the water production rate was independent of the maintenance time. As shown in Figure 7c, the cumulative gas production of Cases 3, 6, and 9 were 85.8×10^4 m^3, 88.9×10^4 m^3, and 90.4×10^4 m^3, respectively. Compared with Case 3, the cumulative gas production of Case 9 increased by 5.36%, indicating that shortening the maintenance time was beneficial to increase the cumulative gas production. From Figure 7d, it can be observed that the cumulative water production of Case 9 (18.8×10^4 m^3) was 5.62% higher than that of Case 3 (17.8×10^4 m^3). The increase in gas production was slightly lower than the increase in water production, which was not the desired result. In summary, shortening the maintenance time is beneficial for improving the gas production rate and cumulative gas production, but it may also result in excessive water production.

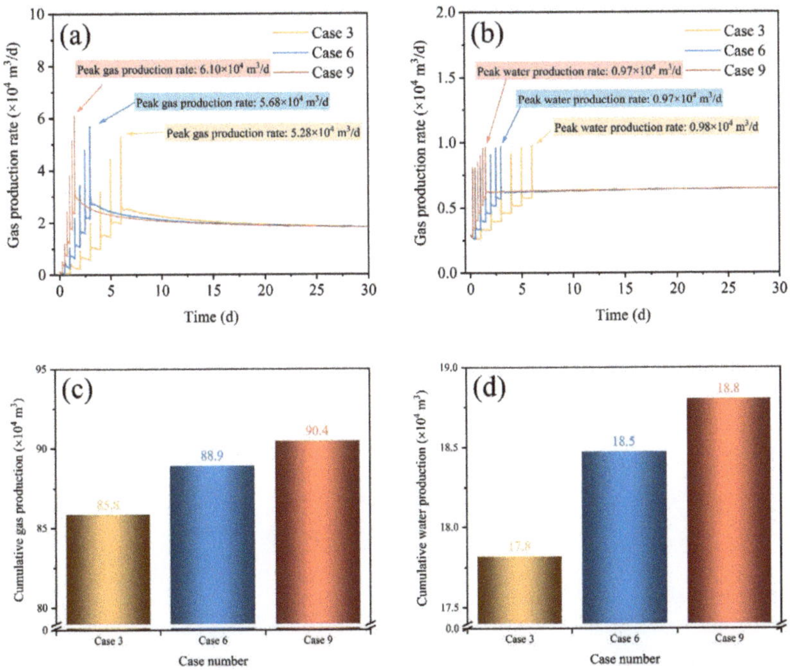

Figure 7. Gas and water production characteristics in Cases 3, 6, and 9: (**a**) gas production rate; (**b**) water production rate; (**c**) cumulative gas production; and (**d**) cumulative water production.

4.3. Comparison of the Cumulative Gas and Water Production for Each Step-Wise Depressurization Pattern

A 100-day simulation was executed to gain a deeper understanding of the effect of the step-wise depressurization method over an extended time frame. The cumulative gas production increment between various step-wise depressurization patterns and the direct depressurization method is depicted in Figure 8. As evidenced, all the step-wise depressurization patterns demonstrated a larger cumulative gas production compared to the direct depressurization method. The cumulative gas production increased as the depressurization gradient decreased and the maintenance time was shortened. The impact of the depressurization gradient on the cumulative gas production surpassed that of the maintenance time on the cumulative gas production. Additionally, it is observed that the cumulative gas production of Case 3 after 100 days was 8.53% higher than that of Case 0, which was even higher than the improvement of 4.63% observed after 30 days (Figure 7). This indicated that the step-wise depressurization method remained effective over a longer period of time. The cumulative water production is shown in Figure 9. It can be observed that the cumulative water production of each case corresponded to its cumulative gas production, with large cumulative gas production being associated with large cumulative water production. Through comparative analyses, it is observed that the increase in cumulative gas production for Case 9 (10.08%) relative to Case 0 was greater than the increase in cumulative water production (8.45%). Through comparison under the same factors, it can be found that compared with Case 1 in terms of depressurization gradient, the cumulative gas production of Case 3 increased by 4.09%, and the cumulative water production increased by 3.10%. Compared with Case 3 in terms of maintenance time, the cumulative gas production of Case 9 increased by 1.43%, and the cumulative water production increased by 1.58%. Therefore, the exploitation conditions used in Case 3 was more favorable for the exploitation of hydrates. In general, step-wise depressurization can alleviate the rapid gas and water production in a short time, and to some extent facilitate the gas exploitation; however, there is still a significant gap between the achieved improvement and commercial exploitation. Thus, it is recommended to combine the step-wise depressurization technique with other methods such as thermal stimulation and permeability enhancement technology for effective exploitation.

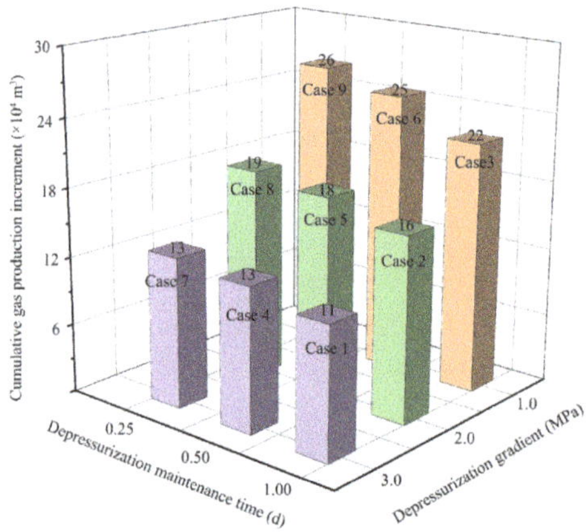

Figure 8. Cumulative gas production increment after 100 days of Cases 1–9.

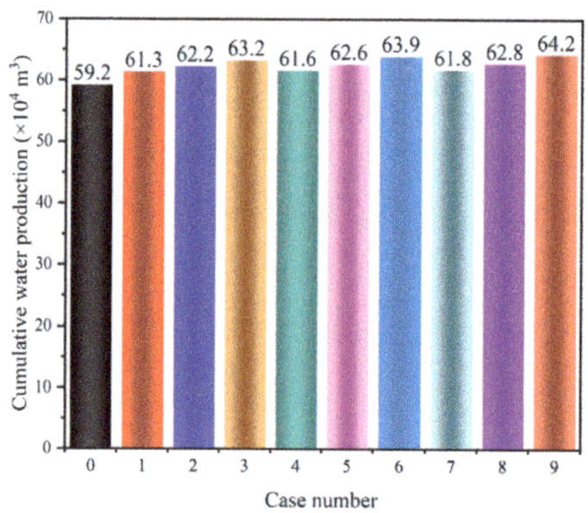

Figure 9. Cumulative water production after 100 days of Cases 0–9.

4.4. Evolution of Reservoir Characteristics Distribution by Step-Wise Depressurization Method

Considering the comprehensive effects of depressurization gradient and maintenance time on gas and water production, as well as the implementation of step-wise depressurization, it was reasonable to select Case 3 as the optimal production condition. Therefore, Case 3 was chosen for presenting the distribution of reservoir characteristics. The evolution of pressure and temperature at various times (6 d and 30 d) were presented in Figure 10. As observed in Figure 10a,b, the pressure decline was mainly concentrated within a range of 50 m, and the propagation speed of pressure in the HBL2 was slower than those in the other two layers. A small area of low temperature was observed near the production well on the sixth day, as shown in Figure 10c, which could be attributed to the rapid decomposition of hydrates and the Joule–Thomson effect [47]. Additionally, it is observed that the temperature in the lower part of the HBL3 began to increase in Figure 10d. Figure 11 revealed the hydrate and gas saturation distributions in the reservoir. It could be observed that the complete decomposition of gas hydrates occurred within 10 m of the production well. The decomposition rate of gas hydrates in the HBL2 was found to be slower than that of the

HBL1 and HBL3, possibly due to the lower permeability of the layer, which hindered the gas–water flow and thus slowed down the decomposition of hydrates. The area of free gas distribution largely overlapped with that of the complete decomposition of gas hydrates. In summary, the evolution of reservoir property distribution by step-wise depressurization exhibited a similar trend to that of direct depressurization [32,41,48]. This similarity could be attributed to the fact that both methods induce gas hydrate decomposition by reducing pressure. However, the step-wise depressurization method yields a larger amount of gas production compared to direct depressurization.

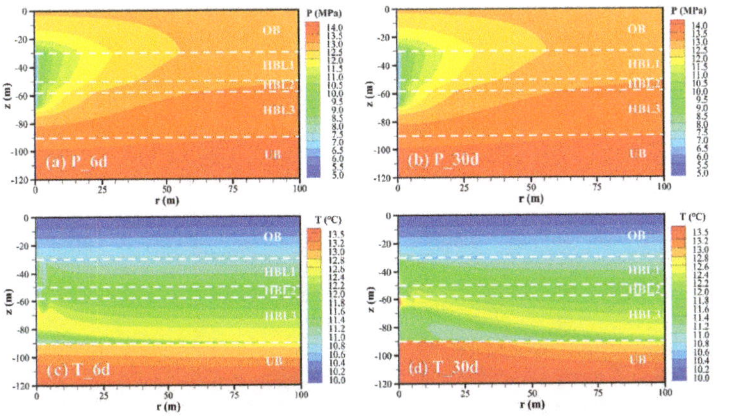

Figure 10. Evolution of the pressure and temperature distributions in the reservoir: (**a**,**b**) pressure distribution at day 6 and day 30; and (**c**,**d**) temperature distribution at day 6 and day 30.

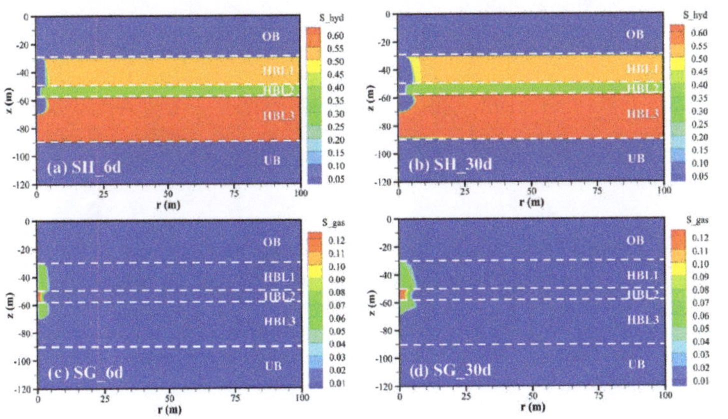

Figure 11. Evolution of the hydrate and gas saturation distributions in the reservoir: (**a**,**b**) gas hydrate saturation distribution at day 6 and day 30; and (**c**,**d**) free gas saturation distribution at day 6 and day 30.

5. Conclusions

In this study, a reservoir model based on the field test results of the Nankai Trough was established, and then a one-year exploitation simulation was conducted for a better understanding of the gas and water production characteristics and the evolution of reservoir characteristics. In addition, the impact of the step-wise depressurization method on the gas extraction from the NGH reservoir was discussed in detail. The main conclusions are as follows:

(1) The effective permeability for the aqueous phase flow will decrease as the decomposition of gas hydrates, while the pore water flow from other layers will eliminate this effect.
(2) The step-wise depressurization method is effective in mitigating short-term excessive gas and water production. A small depressurization gradient and a long maintenance time for each stage can enhance the mitigation effect.
(3) The stepwise depressurization method can increase the cumulative gas production by up to 10% at maximum. Considering the gas and water production characteristics, as well as the difficulty in implementing the step-wise depressurization, it is recommended to adopt a depressurization gradient of 1 MPa and a maintenance time of 1 day.

The simulation results presented in this study primarily focused on evaluating the effectiveness of the step-wise depressurization method for the NGH exploitation in the Nankai Trough of Japan. To provide theoretical support for commercial exploitation, future simulations will be conducted over longer time periods and incorporate various exploitation methods.

Author Contributions: K.X.: Conceptualization, Investigation, Data curation, Writing—original draft; Y.L.: Resources, Funding acquisition, Writing—review and editing; T.Y.: Methodology, Funding acquisition, Writing—review and editing; J.L.: Formal analysis. All authors have read and agreed to the published version of the manuscript.

Funding: This work was supported by the National Natural Science Foundation of China (Grant No. 51976023, 52276053), the National key research and development program (Grant No. 2021YFC2800902) and the Fundamental Research Funds for the Central Universities (Grant No. DUT22LAB115, DUT22RC(3)047).

Data Availability Statement: No new data were created.

Conflicts of Interest: The authors declare that they have no known competing financial interests or personal relationships that could have appeared to influence the work reported in this paper.

References

1. Collett, T.S.; Johnson, A.H.; Knapp, C.C. Natural Gas Hydrates: A Review. *AAPG Mem.* **2009**, *89*, 146–219.
2. Sloan, E.D. Fundamental Principles and Applications of Natural Gas Hydrates. *Nature* **2003**, *426*, 353–359. [CrossRef]
3. Boswell, R. Is Gas Hydrate Energy Within Reach? *Science* **2009**, *325*, 957–958. [CrossRef] [PubMed]
4. Yin, F.; Gao, Y.; Chen, Y.; Sun, B.; Li, S.; Zhao, D. Numerical Investigation on the Long-Term Production Behavior of Horizontal Well at the Gas Hydrate Production Site in South China Sea. *Appl. Energy* **2022**, *311*, 118603. [CrossRef]
5. Lijith, K.P.; Srinivasa Rao, R.; Narain Singh, D. Investigations on the Influence of Wellbore Configuration and Permeability Anisotropy on the Gas Production from a Turbidite Hydrate Reservoir of KG Basin. *Fuel* **2022**, *317*, 123562. [CrossRef]
6. Yang, L.; Guan, D.; Qu, A.; Li, Q.; Ge, Y.; Liang, H.; Dong, H.; Leng, S.; Liu, Y.; Zhang, L.; et al. Thermotactic Habit of Gas Hydrate Growth Enables a Fast Transformation of Melting Ice. *Appl. Energy* **2023**, *331*, 120372. [CrossRef]
7. Collett, T.S. Energy Resource Potential of Natural Gas Hydrates. *AAPG Bull.* **2002**, *86*, 1971–1992. [CrossRef]
8. Li, X.-S.; Xu, C.-G.; Zhang, Y.; Ruan, X.-K.; Li, G.; Wang, Y. Investigation into Gas Production from Natural Gas Hydrate: A Review. *Appl. Energy* **2016**, *172*, 286–322. [CrossRef]
9. Makogon, Y.F.; Holditch, S.A.; Makogon, T.Y. Natural Gas-Hydrates—A Potential Energy Source for the 21st Century. *J. Pet. Sci. Eng.* **2007**, *56*, 14–31. [CrossRef]
10. Gu, Y.; Sun, J.; Qin, F.; Ning, F.; Cao, X.; Liu, T.; Qin, S.; Zhang, L.; Jiang, G. Enhancing Gas Recovery from Natural Gas Hydrate Reservoirs in the Eastern Nankai Trough: Deep Depressurization and Underburden Sealing. *Energy* **2023**, *262*, 125510. [CrossRef]
11. Xue, K.; Liu, Y.; Yu, T.; Yang, L.; Zhao, J.; Song, Y. Numerical Simulation of Gas Hydrate Production in Shenhu Area Using Depressurization: The Effect of Reservoir Permeability Heterogeneity. *Energy* **2023**, *271*, 126948. [CrossRef]
12. Song, Y.; Cheng, C.; Zhao, J.; Zhu, Z.; Liu, W.; Yang, M.; Xue, K. Evaluation of Gas Production from Methane Hydrates Using Depressurization, Thermal Stimulation and Combined Methods. *Appl. Energy* **2015**, *145*, 265–277. [CrossRef]
13. Wang, Y.; Dong, B.; Zhang, L.; Li, W.; Song, Y. Numerical Simulation of CH_4 Recovery from Gas Hydrate Using Gaseous CO_2 Injected into Porous Media. *J. Nat. Gas Sci. Eng.* **2021**, *95*, 104199. [CrossRef]
14. Jin, G.; Xu, T.; Xin, X.; Wei, M.; Liu, C. Numerical Evaluation of the Methane Production from Unconfined Gas Hydrate-Bearing Sediment by Thermal Stimulation and Depressurization in Shenhu Area, South China Sea. *J. Nat. Gas Sci. Eng.* **2016**, *33*, 497–508. [CrossRef]

15. Yu, T.; Guan, G.; Abudula, A.; Yoshida, A.; Wang, D.; Song, Y. Heat-Assisted Production Strategy for Oceanic Methane Hydrate Development in the Nankai Trough, Japan. *J. Pet. Sci. Eng.* **2019**, *174*, 649–662. [CrossRef]
16. Ke, W.; Chen, D. A Short Review on Natural Gas Hydrate, Kinetic Hydrate Inhibitors and Inhibitor Synergists. *Chin. J. Chem. Eng.* **2019**, *27*, 2049–2061. [CrossRef]
17. Hu, Y.; Shi, X.; Li, Q.; Gao, L.; Wu, F.; Xie, G. Effects of a Polyamine Inhibitor on the Microstructure and Macromechanical Properties of Hydrated Shale. *Petroleum* **2022**, *8*, 538–545. [CrossRef]
18. Kan, J.-Y.; Sun, Y.-F.; Dong, B.-C.; Yuan, Q.; Liu, B.; Sun, C.-Y.; Chen, G.-J. Numerical Simulation of Gas Production from Permafrost Hydrate Deposits Enhanced with CO_2/N_2 Injection. *Energy* **2021**, *221*, 119919. [CrossRef]
19. Pandey, J.S.; Ouyang, Q.; von Solms, N. New Insights into the Dissociation of Mixed CH_4/CO_2 Hydrates for CH_4 Production and CO_2 Storage. *Chem. Eng. J.* **2022**, *427*, 131915. [CrossRef]
20. Moridis, G.J.; Silpngarmlert, S.; Reagan, M.T.; Collett, T.; Zhang, K. Gas Production from a Cold, Stratigraphically-Bounded Gas Hydrate Deposit at the Mount Elbert Gas Hydrate Stratigraphic Test Well, Alaska North Slope: Implications of Uncertainties. *Mar. Pet. Geol.* **2011**, *28*, 517–534. [CrossRef]
21. Wang, Y.; Li, X.-S.; Li, G.; Zhang, Y.; Li, B.; Feng, J.-C. A Three-Dimensional Study on Methane Hydrate Decomposition with Different Methods Using Five-Spot Well. *Appl. Energy* **2013**, *112*, 83–92. [CrossRef]
22. Yu, T.; Guan, G.; Abudula, A. Production Performance and Numerical Investigation of the 2017 Offshore Methane Hydrate Production Test in the Nankai Trough of Japan. *Appl. Energy* **2019**, *251*, 113338. [CrossRef]
23. Konno, Y.; Fujii, T.; Sato, A.; Akamine, K.; Naiki, M.; Masuda, Y.; Yamamoto, K.; Nagao, J. Key Findings of the World's First Offshore Methane Hydrate Production Test off the Coast of Japan: Toward Future Commercial Production. *Energy Fuels* **2017**, *31*, 2607–2616. [CrossRef]
24. Yamamoto, K.; Terao, Y.; Fujii, T.; Ikawa, T.; Seki, M.; Matsuzawa, M.; Kanno, T. Operational overview of the first offshore production test of methane hydrates in the Eastern Nankai Trough. In Offshore Technology Conference. *Offshore Technology Conference*. 2014. Available online: https://onepetro.org/OTCONF/proceedings-abstract/14OTC/3-14OTC/D031S034R004/172106 (accessed on 11 June 2023).
25. Ye, J.; Qin, X.; Xie, W.; Lu, H.; Ma, B.; Qiu, H.; Liang, J.; Lu, J.; Kuang, Z.; Lu, C.; et al. The Second Natural Gas Hydrate Production Test in the South China Sea. *China Geol.* **2020**, *3*, 197–209. [CrossRef]
26. Li, J.; Ye, J.; Qin, X.; Qiu, H.; Wu, N.; Lu, H.; Xie, W.; Lu, J.; Peng, F.; Xu, Z.; et al. The First Offshore Natural Gas Hydrate Production Test in South China Sea. *China Geol.* **2018**, *1*, 5–16. [CrossRef]
27. Moridis, G.J.; Reagan, M.T. Estimating the Upper Limit of Gas Production from Class 2 Hydrate Accumulations in the Permafrost: 1. Concepts, System Description, and the Production Base Case. *J. Pet. Sci. Eng.* **2011**, *76*, 194–204. [CrossRef]
28. Heeschen, K.U.; Abendroth, S.; Priegnitz, M.; Spangenberg, E.; Thaler, J.; Schicks, J.M. Gas Production from Methane Hydrate: A Laboratory Simulation of the Multistage Depressurization Test in Mallik, Northwest Territories, Canada. *Energy Fuels* **2016**, *30*, 6210–6219. [CrossRef]
29. Yang, M.; Zheng, J.; Gao, Y.; Ma, Z.; Lv, X.; Song, Y. Dissociation Characteristics of Methane Hydrates in South China Sea Sediments by Depressurization. *Appl. Energy* **2019**, *243*, 266–273. [CrossRef]
30. Zhao, J.; Liu, Y.; Guo, X.; Wei, R.; Yu, T.; Xu, L.; Sun, L.; Yang, L. Gas Production Behavior from Hydrate-Bearing Fine Natural Sediments through Optimized Step-Wise Depressurization. *Appl. Energy* **2020**, *260*, 114275. [CrossRef]
31. Moridis, G.J.; Kowalsky, M.B.; Pruess, K. *User's Manual: A Code for the Simulation of System Behavior in Hydrate-Bearing Geologic Media*; Lawrence Berkeley National Laboratory: Berkeley, CA, USA, 2012; p. 284.
32. Zhu, H.; Xu, T.; Yuan, Y.; Xia, Y.; Xin, X. Numerical Investigation of the Natural Gas Hydrate Production Tests in the Nankai Trough by Incorporating Sand Migration. *Appl. Energy* **2020**, *275*, 115384. [CrossRef]
33. Sun, J.; Ning, F.; Zhang, L.; Liu, T.; Peng, L.; Liu, Z.; Li, C.; Jiang, G. Numerical Simulation on Gas Production from Hydrate Reservoir at the 1st Offshore Test Site in the Eastern Nankai Trough. *J. Nat. Gas Sci. Eng.* **2016**, *30*, 64–76. [CrossRef]
34. Buković, D.; Carek, V.; Durek, D.; Kuna, T.; Keros, J. Measurement of Magnetic Field in Dentistry. *Coll. Antropol.* **2000**, *24* (Suppl. S1), 85–89. [PubMed]
35. Song, H.B.; Jiang, W.W.; Zhang, W.S.; Hao, T. Progress on Marine Geophysical Studies of Gas Hydrates. *Prog. Geophys.* **2002**, *17*, 224–229.
36. Yu, T.; Guan, G.; Abudula, A.; Yoshida, A.; Wang, D.; Song, Y. Application of Horizontal Wells to the Oceanic Methane Hydrate Production in the Nankai Trough, Japan. *J. Nat. Gas Sci. Eng.* **2019**, *62*, 113–131. [CrossRef]
37. Feng, Y.; Chen, L.; Suzuki, A.; Kogawa, T.; Okajima, J.; Komiya, A.; Maruyama, S. Numerical Analysis of Gas Production from Layered Methane Hydrate Reservoirs by Depressurization. *Energy* **2019**, *166*, 1106–1119. [CrossRef]
38. A Closed-Form Equation for Predicting the Hydraulic Conductivity of Unsaturated Soils. Available online: https://acsess.onlinelibrary.wiley.com/doi/epdf/10.2136/sssaj1980.03615995004400050002x (accessed on 23 March 2022).
39. Van Genuchten, M.T. A closed-form equation for predicting the hydraulic conductivity of unsaturated soils. *Soil Sci. Soc. Am. J.* **1980**, *44*, 892–898. [CrossRef]
40. Chen, L.; Feng, Y.; Kogawa, T.; Okajima, J.; Komiya, A.; Maruyama, S. Construction and Simulation of Reservoir Scale Layered Model for Production and Utilization of Methane Hydrate: The Case of Nankai Trough Japan. *Energy* **2018**, *143*, 128–140. [CrossRef]

41. Mao, P.; Sun, J.; Ning, F.; Chen, L.; Wan, Y.; Hu, G.; Wu, N. Numerical Simulation on Gas Production from Inclined Layered Methane Hydrate Reservoirs in the Nankai Trough: A Case Study. *Energy Rep.* **2021**, *7*, 8608–8623. [CrossRef]
42. Merey, S.; Chen, L. Numerical Comparison of Different Well Configurations in the Conditions of the 2020-Gas Hydrate Production Test in the Shenhu Area. *Upstream Oil Gas Technol.* **2022**, *9*, 100073. [CrossRef]
43. Yu, T.; Guan, G.; Abudula, A.; Wang, D. 3D Visualization of Methane Hydrate Production Behaviors under Actual Wellbore Conditions. *J. Pet. Sci. Eng.* **2020**, *185*, 106645. [CrossRef]
44. Yu, T.; Chen, B.; Jiang, L.; Zhang, L.; Yang, L.; Yang, M.; Song, Y.; Abudula, A. Feasibility Evaluation of a New Approach of Seawater Flooding for Offshore Natural Gas Hydrate Exploitation. *Energy Fuels* **2023**, *37*, 4349–4364. [CrossRef]
45. Wei, R.; Xia, Y.; Wang, Z.; Li, Q.; Lv, X.; Leng, S.; Zhang, L.; Zhang, Y.; Xiao, B.; Yang, S.; et al. Long-Term Numerical Simulation of a Joint Production of Gas Hydrate and Underlying Shallow Gas through Dual Horizontal Wells in the South China Sea. *Appl. Energy* **2022**, *320*, 119235. [CrossRef]
46. Cao, X.; Sun, J.; Ning, F.; Zhang, H.; Wu, N.; Yu, Y. Numerical Analysis on Gas Production from Heterogeneous Hydrate System in Shenhu Area by Depressurizing: Effects of Hydrate-Free Interlayers. *J. Nat. Gas Sci. Eng.* **2022**, *101*, 104504. [CrossRef]
47. Yarveicy, H.; Ghiasi, M.M.; Mohammadi, A.H. Determination of the Gas Hydrate Formation Limits to Isenthalpic Joule–Thomson Expansions. *Chem. Eng. Res. Des.* **2018**, *132*, 208–214. [CrossRef]
48. Feng, Y.; Chen, L.; Merey, S.; Lijith, K.P.; Singh, D.N.; Komiya, A.; Maruyama, S. Numerical Modelling of Gas Production from the Oceanic Gas Hydrate Reservoirs in Eastern Nankai Trough (AT1 Site), Japan. *Environ. Geotech.* **2020**, *10*, 176–185. [CrossRef]

processes

MDPI

Article

Effect of Different Concentrations of NiMnGa Micro/Nanoparticles on the Kinetics of Natural Gas Hydration

Zhiwei Zhao [1], Qiong Wu [1,2], Zhen Li [2,*], Huiyuan Meng [1], Maged Elhefnawey [2,3], Xinyan Wang [1], Qiang Wu [1], Li Li [2,4] and Baoyong Zhang [1,*]

1 Department of Safety Engineering, Heilongjiang University of Science and Technology, Harbin 150022, China
2 International Joint Laboratory of Advanced Nanomaterials of Heilongjiang Province (International Cooperation), Harbin 150001, China
3 Department of Mechanical Engineering, Faculty of Engineering, Kafrelsheikh University, Kafrelsheikh 33156, Egypt
4 International Joint Laboratory of Advanced Bulk Nanomaterials for Innovative Applications, Harbin Engineering University, Harbin 150001, China
* Correspondence: lz_heu@hrbeu.edu.cn (Z.L.); byzhang1982@163.com (B.Z.)

Abstract: To improve gas hydrate storage and transportation technology, ferromagnetic intermetallic compound NiMnGa particles with martensitic transformation endothermics were used to form micro/nanofluids with sodium dodecyl sulfate (SDS) to further strengthen the gas hydration process. In this work, the kinetic process of gas hydration in NiMnGa fluids with different concentrations (0, 0.1, 1, 2, and 3 wt.%) was investigated using a rotating magnetic field gas hydration separation experimental setup. The results show that the induction time of the 3 wt.% NiMnGa system was shortened by 98.3%, the gas consumption was increased by 50.5%, and the gas consumption rate was increased by 351.9% compared with the SDS system. Therefore, it is inferred from the mass transfer that NiMnGa micro/nanofluids can accelerate the formation of hydrates.

Keywords: natural gas hydrate; NiMnGa; micro/nanoparticles; particle concentration; kinetic characteristics

1. Introduction

Natural gas hydrates are a widely distributed clean energy source [1,2]. With the successful test mining of marine sediment hydrates in China, gas hydrates are expected to become a new generation of energy to displace oil, coal, and conventional gas. The development and utilization of gas hydrates has become essential to ensuring national energy security, promoting deep sea strategy, and achieving the double carbon goals. Gas hydrates are a non-chemometric cage-like crystalline substance formed by water and methane gas, known as combustible ice [3,4], making it a very clean energy source. Based on the advantages of mild formation conditions, high gas content, and safe storage and transportation [5,6], related scholars have proposed the safe storage and transportation of methane gas released via hydrate mining through hydration technology so as to alleviate energy tension [7–12]. However, there is a problem in the hydration process, namely, the mass transfer, which has long induction periods, slow formation rates, and high driving forces of hydrate formation [13].

Improving the kinetics of hydrate generation is one of the feasible factors for the industrialization of hydrate technology. Scholars have proposed many methods by which to improve the kinetics of hydrate generation through experimental studies, and the traditional chemical methods are mainly intended to reduce the surface tension of the liquid phase by adding kinetic enhancers to promote the formation of hydrates [14,15]. The typical kinetic additives mainly include SDS [16,17], amino acids [18,19], and sulfonated lignin (SL) [20]. The traditional physical hydration enhancement methods mainly include the stirring method, spraying method, bubbling method, and applied magnetic field method. In addition, some scholars have found that the properties of nanoparticles, such as high specific surface area and high thermal conductivity, can effectively enhance the heat and mass transfer in the hydration process [21–24], which can improve kinetic parameters, such as the hydrate conversion rate and gas storage, and promote hydrate generation.

Ghozatloo et al. [25] first found that 1 wt.% graphene reduced the induction time of 92.6% CH_4 gas hydrate by 61.07% and increased the storage capacity of the hydrate by 12.9%. Park et al. [26] used different concentrations (0.001–0.006 wt.%) of carbon nanotubes (MWCNTs) to improve CH_4 hydrate generation dynamics, where 0.004 wt.% of multi-walled carbon nanotubes resulted in milder phase equilibrium conditions for CH_4 hydrate, about 300% higher gas consumption than pure water, and a significant reduction in the CH_4 hydrate formation time. Arjang et al. [27] used a mixed solution containing sodium citrate and nanosilver to improve hydrate generation kinetics. Compared to sodium citrate solution, the induction time of the mixed solution with nanosilver was reduced by 63.6%, and the CH_4 gas consumption was increased by 7.4%. Rahmati-Abkenar et al. [28,29] discovered that an aqueous solution of Ag nanoparticles shortens the induction time of CH_4 hydrate by 97% compared with pure water, and that increasing the particle concentration can reduce the effect of surface tension on CH_4 hydrate formation. Aliabadi et al. [30] investigated the effect of 0.035 wt.% SDS and 1 wt.% CuO nanoparticles on methane hydrate formation, the results showed that the addition of nanoparticles reduced the induction time and increased the final hydrate conversion by 157.93% compared to pure water. Abdi-Khanghah et al. [31] conducted 0.05 and 0.1 wt.% ZnO + 0.03 wt.% SDS particle experiments on methane hydrate, wherein ZnO particles improved heat and mass transfer in the solution due to their large surface area and nucleation sites. The results showed that the addition of nanoparticles decreased the induction time of the hydration reaction and increased the hydrate conversion by 60.22%. Kakati et al. [32] conducted experiments on the kinetics of $CH_4 + C_2H_6 + C_3H_8$ hydrate formation in 0.1, 0.4, and 0.8 wt.% Al_2O_3/ZnO + 0.03 wt.% SDS solution nanofluids and found that the incorporation of nanoparticles increased the gas consumption, which was 121% higher than that of the pure water system, and the gas storage. Meanwhile, some scholars have treated the nanoparticles by means of grafting functional groups and solid-loaded coated surfactants to promote the kinetics of hydrate generation. Wang et al. [33] realized SDS-coated nanospheres (SDS@PSNS) and nano-Ag particles grafted on the surface of SDS@PSNS to prepare Ag&SDS-coated nanospheres (Ag&SDS@PSNS) in order to enhance its promoting effect on the formation of methane hydrates. The hydrate formation time with 0.1 mol/L Ag&SDS@PSNS was shortened from 422.7 min to 156.7–273 min, and the entire methane hydrate generation period can be completed within 50 min with the 0.2 mmol/L Ag&SDS@PSNS. Chen et al. [34] investigated the synthesis of SDS-coated Fe_3O_4 nanoparticles that were successfully used to promote hydrate formation and cycling experiments, and SDS-coated Fe_3O_4 promoted hydrate formation more effectively than bare Fe_3O_4. With the 20 g/L SDS-coated Fe_3O_4, the CH_4 storage was 130 ± 5.9 v/v, and the induction and reaction times were reduced to 77.6 ± 24.1 and 36 ± 3 min, respectively.

All of the above nanofluids belong to the category of non-phase-change nanofluids, and studies have shown that nanoparticles have a promotional effect on hydrate generation, and the promotional effect varies depending on the concentration, type, and size of nanoparticles. In addition, it is expected to further strengthen the promotional effect of nanofluids on hydrate generation by utilizing the heat-absorbing property of materials in the process of phase change to absorb the heat of hydrate generation. Yan et al. [35] investigated the solid-state storage, generation rate, and induction period of CH_4 hydrates in nTD slurry and found that the hydration rate was highest under the conditions of 5.26 MPa, 158.7 v/v conditions with the highest hydration rate, and a hydration time of only 78 min. Song et al. [36] investigated the growth rate of CH_4 hydrates in cement slurries containing 25–45 wt.% solid n-tetradecane in the range of pressures of 4.7–6.46 MPa and temperatures of 263.2–283.2 K. The growth rates of the hydrates obtained using phase change material particles were increased by a factor of three to nine. Liu et al. [37] demonstrated that under the action of a specific magnetic field, the amount of hydrate formation significantly increases, the hydration rate can reach 100%, and the induction time can be greatly shortened from 9 h to 40 min; moreover, the polarity of the magnetic field, the strength of the magnetic field, and the number of iron wires all have an impact on the formation of hydrates. Sun et al. [38] found that the magnetic effect increases the nucleation rate of CO_2 hydrates, shortens the nucleation induction time, and increases the consumption and rate of CO_2. This is mainly due to the magnetic field reducing the contact angle between the water and the solid surface, reducing the size and free energy of hydrate nucleation, thereby promoting hydrate nucleation.

At present, related studies at home and abroad have focused on the process of gas hydrate formation in non-phase transformation nanofluids. Ferromagnetic intermetallic compound NiMnGa micro/nanomaterials, first, have the characteristics of large specific surface area, small size, and magnetic characteristic [39]. Second, NiMnGa has phase transformation characteristics, can absorb heat when reverse martensitic transformation occurs, and can provide a good environment for hydrate growth, thus promoting hydrate formation [40,41]. In addition, the temperature and latent heat of NiMnGa phase transformation can be controlled in the experimental temperature range easily [42]. In addition, a peripheral rotating magnetic field is used in this work, which can disperse magnetic micro/nanoparticles in the solution and promote mass transfer. Therefore, in this paper, the NiMnGa alloy particles, which has magnetic, high thermal conductivity, and phase transformation heat absorption characteristics, is used to improve the mass and heat transfer constraints of the hydration process for the purpose of shortening the induction time and improving the gas storage capacity of hydrate formation [43].

2. Materials and Methods

2.1. Experimental Materials

High-purity nickel blocks (99.99%), electrolytic manganese (99.7%), and high-purity gallium (99.99%) were supplied by Zhongnuo New Materials (Beijing, China) Technology Co. Gas samples (G1: 60% CH_4, 32% N_2, 8% O_2; G2: 70% CH_4, 24% N_2, 6% O_2; G3: 80% CH_4, 16% N_2, 4% O_2) were supplied by Harbin Chunlin Gas Distribution Co., Ltd. (Harbin, China); SDS was supplied by Tianjin Bailunsi Biotechnology Co., Ltd. (Tianjin, China); pure water was home-made by laboratory Thermo Fisher Water Purifier (Waltham, MA, USA).

2.2. Experimental Apparatus

Based on the requirements of the NiMnGa for the experimental study of gas hydration separation kinetics, a set of rotating magnetic field gas hydration separation experimental apparatus, developed by the group, was used. The experimental and test system is shown in Figure 1. The high-pressure titanium alloy reaction kettle is equipped with two high magnetic steels on the outside, and controlling the spacing between the magnetic steels can generate a magnetic field in the range of 0~0.5 T. The magnetic steel is fixed on a rotating electric platform, and the rotation of the magnetic field causes the magnetic particles to move uniformly and stably in the solution inside the reactor with a speed range of 0–200 r/min. A detailed introduction of the device can be found in reference [41].

2.3. Experimental Procedure

Based on the extensive research on material preparation process optimization conducted by the research group in the early stage, this article uses the ball milling method to prepare NiMnGa alloy particles, and the preparation steps are as follows [44–49]: (1) in a vacuum arc melting furnace, nickel blocks, manganese, and gallium are melted into alloy ingots in the ratio of $Ni_{52.5}Mn_{22.5}Ga_{25}$; in order to supplement the volatilization loss of Mn during the melting process, 5% Mn is added to the ingredient; (2) we perform homogenization and solid solution treatment on alloy ingots; (3) NiMnGa ingots are milled in a high-speed vibrating ball mill to obtain NiMnGa particles; and (4) we anneal the particles after ball milling. The NiMnGa-SDS micro/nanofluid was prepared by using NiMnGa micro/nanoparticles and SDS solution.

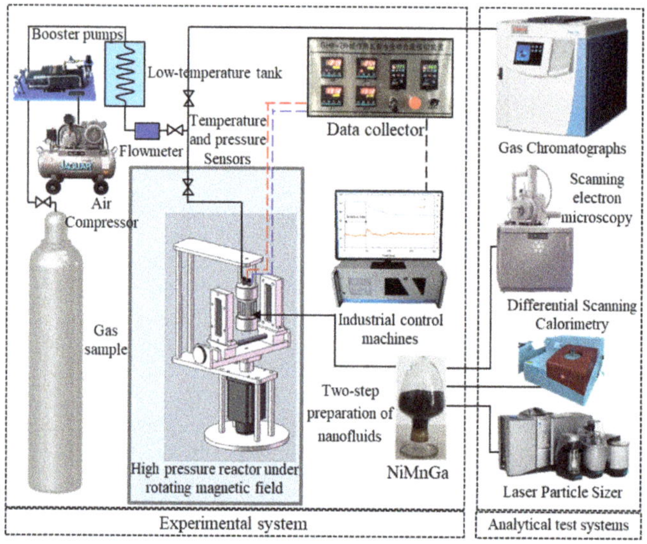

Figure 1. Rotating magnetic field gas hydration separation experimental setup.

The fluid Is prepared by using NiMnGa particles, SDS, and pure water. We place SDS and particles in 40 mL pure water (prepared by a secondary pure water analyzer, Pacific TII, Thermo Fisher, Waltham, MA, USA). Then, a dispersing machine (T18 digital TURRAX, IKA, Staufenim, Germany) was used to stir and disperse the solution in 12,000 rad/min for 90 s, and the treated solution was loaded into the reactor for later use. The detailed process of fluid preparation and the hydration experiment can be found in reference [41].

2.4. Determination of Kinetic Parameters

Using graphical method, the induction time is defined as the time required from the initial pressure reached to the first significant temperature increase and pressure decrease during the hydration reaction process.

Gas consumption Δn:

$$\Delta n = \frac{n_0 - n_t}{n_w} = \frac{\frac{P_0 v_0}{z_0 RT_0} - \frac{p_t v_t}{z_t RT_t}}{n_w}. \quad (1)$$

In the formula, n_0 and n_t represent the amount of gas substances in the reactor at the initial time and time t, respectively; n_w is the molar amount of water; P_0 and T_0 are the initial pressure and temperature values in the reactor, respectively; P_t and T_t are the pressure and temperature values in the reactor at time t during the reaction process, respectively; R is the gas constant; and Z_0 and Z_t are the compression factors at the initial time and time t, respectively.

Gas consumption rate v:

$$v = \left(\frac{d\Delta n}{dt}\right)_t \approx \frac{\Delta n}{dt}. \quad (2)$$

In the formula, t is the time difference between two certain moments.

3. Results

For phase transformation testing and characterisation of NiMnGa micro/nanoparticles using high-pressure differential scanning calorimetry, scanning clicks, and particle size analysis via a laser particle sizer, the results of determination, characterisation, and analysis are given in reference [48].

In order to optimize the mass transfer effect of NiMnGa micro/nanofluids on the hydration process of natural gas, a vertical rotating magnetic field was set outside the titanium alloy high-pressure reactor used in this paper. The magnetic field rotation drives the NiMnGa alloy particles in the reactor to flow in the liquid phase, and the purpose of the uniform dispersion of particles is achieved by adjusting the magnetic field size and rotation speed. Under the premise of energy saving, it is found that the minimum rotational speed and the minimum magnetic field that can make the particles flow uniformly and disperse uniformly in the liquid phase are 100 rpm and 0.09 T, respectively. In this paper, the magnetic properties of NiMnGa alloy particles were tested via a physical property measurement system (PPMS-9, Quantum Design) under magnetic fields of 0 T and 0.09 T, respectively. The test temperature control speed was 5 °C/min, the magnetic field resolution was 0.2 mT, and the temperature range was −50~120 °C.

The gas hydrate formation process was conducted at 6.2 MPa and 2 °C. The experiments were carried out in the SDS system (NiMnGa particle Conc. 0%) and the NiMnGa-SDS system with different concentrations (NiMnGa particle Conc. 0.1, 1, 2, and 3 wt.%) using three different gas samples (G1, G2, G3). Each experiment was repeated three times, and the results were averaged three times. Table 1 displays the average experimental results for gas hydration in the SDS and NiMnGa-SDS systems. In addition, the Supporting Information Table S1 displays all the experimental results for gas hydration in the SDS and NiMnGa-SDS systems. Figure 2 depicts the temperature and pressure profiles from the end of gas injection to hydrate formation stabilization. The process of producing gas hydrates is divided into four stages: nucleation period (ab); rapid hydrate growth period (bc); slow hydrate growth period (cd); and stabilization period (de).

Table 1. The average results for the gas hydration kinetics experiment.

Particle Conc. (wt.%)	Exp. No.	Induction Time (min)	Gas Consumption (mmol/mol)	Hydration Reaction Time (min)	Gas Consumption Rate (mmol/mol/min)
0	G1-1	113.9	9.7	185.6	0.054
	G2-1	48.3	16.5	155.9	0.113
	G3-1	42.9	23.7	107.1	0.222
0.1	G1-2	3.4	11.9	92.2	0.135
	G2-2	8.9	16.6	76.4	0.221
	G3-2	5.8	26.2	74.0	0.360
1	G1-3	4.0	13.2	70.9	0.196
	G2-3	2.4	21.6	82.1	0.264
	G3-3	1.4	25.0	79.6	0.312
2	G1-4	5.3	10.8	48.0	0.229
	G2-4	2.6	21.4	59.0	0.363
	G3-4	3.0	27.1	67.0	0.410
3	G1-5	1.9	14.6	59.9	0.244
	G2-5	1.3	19.0	57.2	0.335
	G3-5	0.9	27.0	61.1	0.444

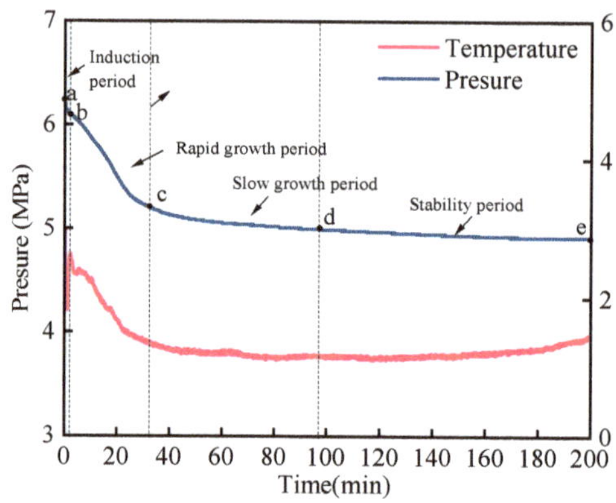

Figure 2. Temperature and pressure curves of hydrate formation in the G1-3 NiMnGa system.

4. Discussion

4.1. The Magnetic Properties of NiMnGa Alloy Particles

Figure 3 depicts the magnetization–temperature curves of NiMnGa alloy particles under 0 T and 0.09 T conditions. In the 0 T magnetic field, the magnetization is close to 0.4 emu/g. The difference of magnetization is about 30 emu/g at 0.09 T, which is much larger than that at 0 T, indicating that the increase in magnetic field can improve the magnetization of particles. At the same time, it also shows that NiMnGa alloy particles are magnetic at 0.09 T and can rotate in the solution with a 0.09 T magnetic field during the hydration reaction of natural gas to form micro/nanofluids.

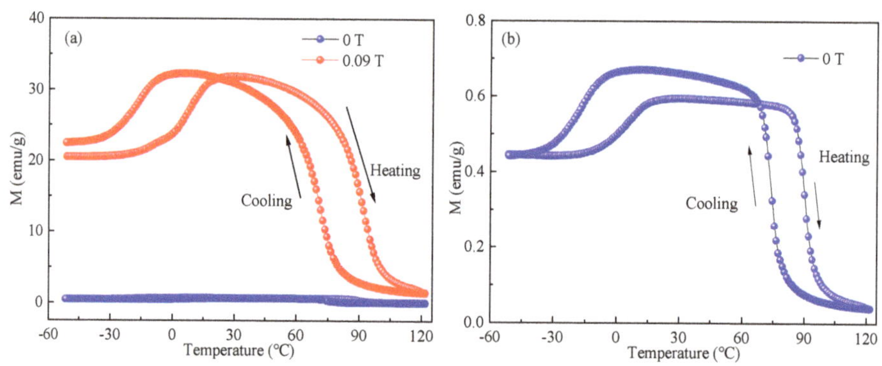

Figure 3. The magnetization–temperature curves of NiMnGa particles: (**a**) magnetic fields in 0 T and 0.09 T; (**b**) the enlarged view in 0 T.

4.2. The Effect of NiMnGa Particle Concentration

4.2.1. The Induction Time

Induction time is a critical parameter in hydrate kinetics because it determines how quickly hydrate nucleation occurs. Figure 4 shows the variation of hydrate induction time in different gas samples with NiMnGa content. The induction time of all the gas samples is shortened greatly in the NiMnGa-SDS system as compared to the SDS system. With the increasing content of NiMnGa particles, the induction time further decreases slightly. The average induction time of the 0.1 wt.% NiMnGa-SDS system in three gas samples was 3.4, 8.88, 5.83 min, respectively; compared with the SDS system, the average induction time was shortened by about 97.0%, 81.6%, and 86.4%, respectively. It can be seen that the addition of only 0.1 wt.% NiMnGa can shorten the induction time of hydrates significantly in different gas samples.

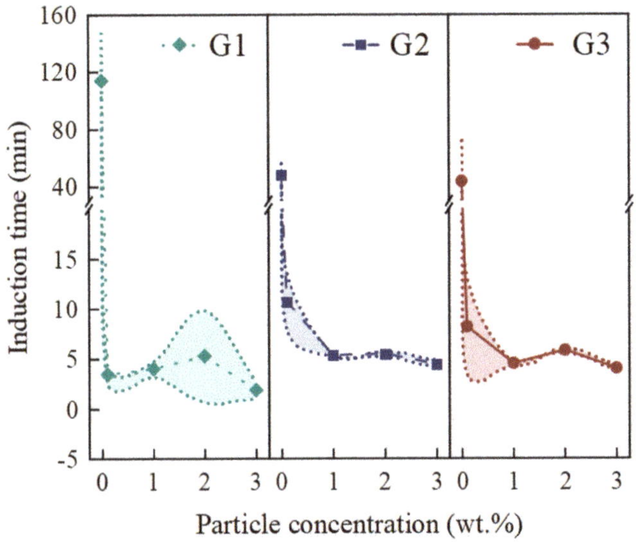

Figure 4. Induction time versus particle concentration curves.

Furthermore, in different concentrations of the NiMnGa-SDS system, the induction time of adding 3 wt.% NiMnGa was the shortest, which is 1.9, 1.3, and 0.9 min in the G1, G2, and G3 gas samples, respectively. These values of induction time are shortened by 98.3%, 97.2%, and 98.0% compared with the SDS system. As compared with SDS system, the small particle size of NiMnGa provides more nucleation sites. Meanwhile, particles collide at the gas–liquid interface to increase gas transfer due to Brownian motion [41]. Therefore, the induction time is shortened significantly by a small amount of NiMnGa particles.

On the other hand, in the SDS system, the induction time of the three gas hydrates decreases with the increase in CH_4 content; that is, the increase in CH_4 content contributes to the initial nucleation of hydrates in the absence of particles. However, in the NiMnGa-SDS system, the effect of CH_4 content on the induction time is not monotonous, which is different from the SDS system, but the three are concentrated between 0.9–9 min, indicating that NiMnGa particles are suitable for different CH_4 content gases in shortening the induction time.

4.2.2. The Gas Consumption

Figure 5 shows the dependence of gas consumption on particle content. The gas consumption of the NiMnGa-SDS system is greater than that of the SDS system and exhibits an increasing trend with the increase in particle concentration, although the effect of particle content is not identical for three gas samples. For example, the average gas consumption of the 0.1 wt.% NiMnGa-SDS system was 11.9, 16.6, and 26.2 mmol/mol, respectively, which was 22.4%, 0.5%, and 10.5% higher than that of the SDS system. In the G1 system, the gas consumption of the 3 wt.% NiMnGa-SDS system is the largest (14.6 mmol/mol), which is about 50.5% than that of the SDS system. With the increase in NiMnGa concentration, the metal surface that can provide hydrate formation increases. At the same time, NiMnGa particles move with the rotation of the magnetic field, and the collision probability of high-concentration particles increases at the gas–liquid interface and in the liquid phase, so that more CH_4 molecules can enter the liquid phase and begin to form on more metal surfaces, thereby increasing gas consumption.

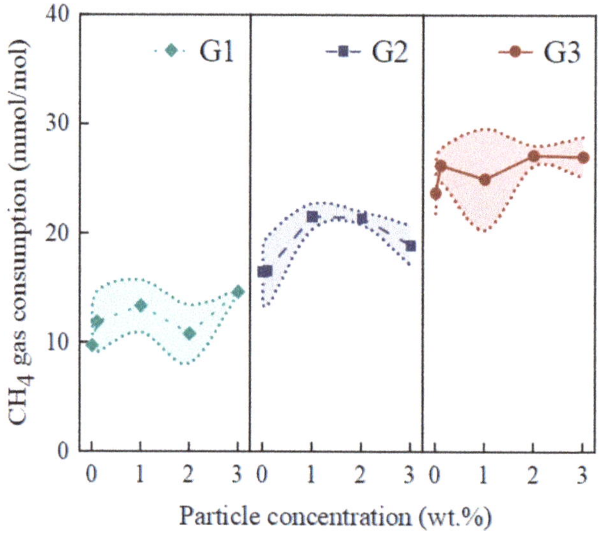

Figure 5. Gas consumption versus particle concentration curves.

In addition, whether in the SDS system or the NiMnGa-SDS system, the gas consumption of the three gas hydrates increases with the increase in CH_4 content. It can also be seen in Figure 4 that in the G3 gas sample, the gas consumption of the SDS system alone is higher than that of all the systems in the G2 gas sample, and the G2 system is also higher than the G1 system. This indicates that although the gas consumption is increased by the NiMnGa particles with the concentration range used in the experiment, the CH_4 content is still the primary factor determining the amount of hydrate formation.

4.2.3. The Gas Consumption Rate

Figure 6a,b depicts the reaction time and gas consumption rate in various gas samples. In Figure 6a, all the hydrate reaction times of the NiMnGa-SDS system are less than 93 min, which is much shorter than those of the SDS system. It shows that the hydration reaction time can be shortened by NiMnGa particles effectively, and with the increase in NiMnGa particle concentration in the three gas samples, the reaction time of the hydrates decreases as a whole. Similar to the induction time, in the SDS system, the reaction time of the three gas hydrates decreases with the increase in methane content. However, in the NiMnGa-SDS system, the reaction time of all three gases is concentrated between 48–92 min, and the extent of shortening the reaction time of gas hydrates with low-concentration methane content (G1) is greater than that of gas hydrates with high-concentration methane content (G3), indicating that NiMnGa particles have a more significant effect on gas hydrates with lower-concentration methane content in shortening the reaction time.

Correspondingly, in Figure 6b, the gas consumption rate of the NiMnGa-SDS system is larger than that of the SDS system and exhibits an increasing trend with the increase in CH_4 concentration and particle concentration. The average gas consumption rate of the 0.1 wt.% NiMnGa-SDS system in the G3 gas sample is 0.360 mmol/mol/min, which is about 166.7% and 62.9% higher than the average gas consumption rate of the same NiMnGa concentration system in the G1 and G2 gas samples (0.135, 0.221 mmol/mol/min).

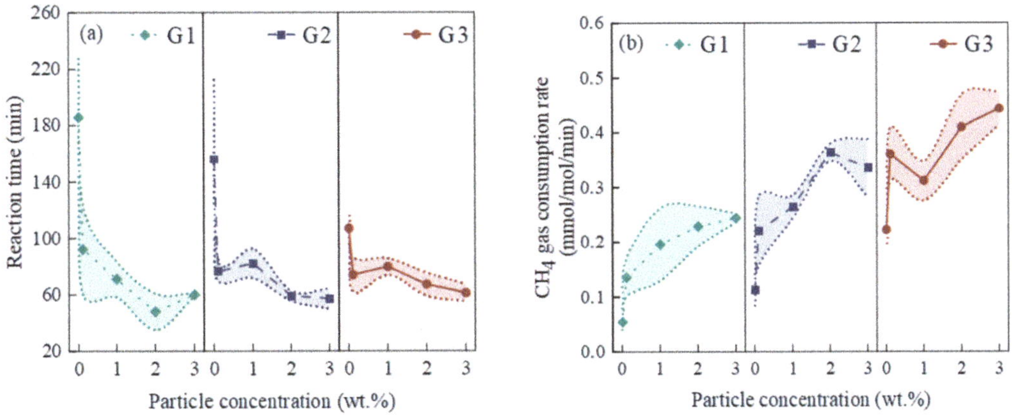

Figure 6. Curves of gas consumption rate with particle concentration: (**a**) reaction time; (**b**) gas consumption rate.

At the same time, adding a small amount of NiMnGa alloy particles has different degrees of improvement in the gas consumption rate among the three gas samples. The average gas consumption rates of the 0.1 wt.% NiMnGa-SDS system were 0.135, 0.221, and 0.36 mmol/mol/min, respectively, which were 150%, 95.6%, and 62.2% higher than those of the SDS system. In addition, the gas consumption rate for 3% NiMnGa in the G1 system is 0.244 mmol/mol/min, which is 351.9%, 80.7%, 24.5%, and 6.6% higher than that of 0 wt%, 0.1 wt.%, 1 wt.%, and 2 wt.% NiMnGa-SDS systems. This indicates that the higher concentration of NiMnGa is conducive to the rapid formation of hydrates in the experimental range of this paper. NiMnGa particles increase the gas–liquid reaction interface so that more CH_4 hydrates can grow at the same time. The gas consumption rate increases in a high speed during the rapid growth period of hydrates. The higher concentration of NiMnGa realizes the transport of more NiMnGa particles to the gas, and more CH_4 gas molecules enter the liquid phase continuously and participate in the formation of hydrates. Therefore, with the increase in NiMnGa concentration, the gas consumption rate has different degrees of upward trend.

4.3. Analysis of Gas Hydrate Enhancement by NiMnGa Micro/Nanofluids

4.3.1. The Effect of NiMnGa Particle Concentration on the Kinetics of the Gas Hydration Reaction

The magnetic NiMnGa particles are combined with SDS solution to form micro/nanofluids by rotating magnetic field, to accelerate the kinetic process of the natural gas hydration reaction. Figure 7 shows the gas hydration process and the enhanced effect of gas hydration via different concentrations of NiMnGa micro/nanofluids. The NiMnGa-SDS system can effectively improve the hydrate kinetic conditions in different gas systems. The small particle size of NiMnGa micro/nanofluids can provide more reaction interfaces, increase hydrate nucleation and formation rates, increase heat exchange area between particles and fluid, and accelerate heat transfer during the hydration reaction. At the same time, due to Brownian motion, the particles collide at the gas–liquid interface to achieve gas transport purposes. The number of micro/nanoparticles per unit volume increases as the NiMnGa concentration increases, increasing the chance of inter-particle collision and increasing the efficiency of material transfer and heat transfer, while more NiMnGa particles can carry out gas transport at the gas–liquid interface and play a role in mass transfer.

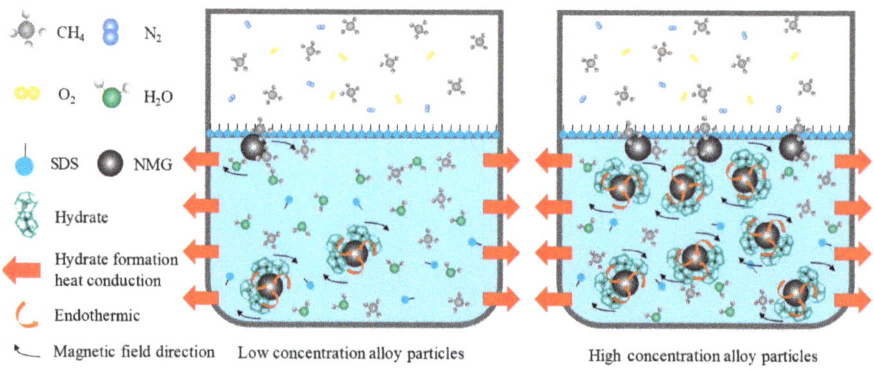

Figure 7. Mechanism of CH_4 hydrate enhancement via different concentrations of NiMnGa.

In addition, NiMnGa phase transformation alloy is a thermoelastic martensitic transformation shape memory material [49]. When the hydrates are formed rapidly, NiMnGa particles can absorb part of the hydrate formation heat to reduce the ambient temperature appropriately [48], thereby promoting the hydrate formation process.

4.3.2. The Effect of External Magnetic Field on the Kinetics of Gas Hydration Reaction in NiMnGa Micro/Nanofluids

Combined with the magnetic properties of NiMnGa alloy, an experimental system of gas hydration reaction in a dynamic rotating magnetic field was built in this paper. The system is equipped with a vertical rotating magnetic field outside the titanium alloy high-pressure reactor, which drives NiMnGa alloy particles inside the reactor to flow in the liquid phase, achieving the strengthening of natural gas hydration reaction kinetics with NiMnGa alloy particles. The rotating magnetic fields can provide a magnetic field strength of 0~0.5 T. Due to the magnetic field, reducing the contact angle between the water and the solid surface, reducing the size and free energy of hydrate nucleation, and enhancing the movement of ions in the gaps between hydrate particles, the magnetic effect promoted the kinetics of natural gas hydration [37,38]. At the same time, the NiMnGa alloy particles used in the experiment, as measured in Section 4.1, exhibit magnetism at a magnetic field strength of 0.09 T and will rotate and flow with a 0.09 T magnetic field during the natural gas hydration reaction to form micro/nanofluids. This transforms the liquid phase into a liquid–solid two-phase suspension, changing the structure of a single liquid phase and promoting the energy transfer process within the micro/nanofluid. At the same time, the rotation of the magnetic field drives the alloy particles to play a strong stirring role in the liquid phase so that the NiMnGa alloy particles are uniformly dispersed and flowing in the base liquid, reducing the possibility of agglomeration. This also accelerates the gas transport of particles in the gas–liquid phase, allowing more gas molecules to enter the liquid phase to participate in the reaction. In the liquid phase, it accelerates the collision rate between particles and between particles and the gas–liquid phase, further promoting the transfer of gas–liquid substances, thus strengthening the dynamic process of natural gas hydration reaction.

5. Conclusions

The micro/nanofluids containing magnetic NiMnGa particles are used to improve the gas hydration process in this paper. Using this fluid system in an external rotating magnetic field, experiments on gas hydrate formation at various NiMnGa concentrations were carried out. The results show that NiMnGa micro/nanoparticles can promote the kinetic process of natural gas hydration reaction. At the same time, this promotion effect is further improved with the increase in NiMnGa concentration between 0.1–3 wt.%, and the promotion effect on the growth process of CH_4 hydrates is the best when the NiMnGa concentration is

3 wt.%. Compared with the SDS system in the G1 system, the induction time of the 3 wt.% NiMnGa system was shortened by 98.3%, the gas consumption was increased by 50.5%, and the gas consumption rate was increased by 351.9%. On the other hand, in different gas sample systems, the optimization of hydrate kinetic parameters becomes more obvious with the increase in CH_4 concentration in the gas samples. In the NiMnGa-SDS system, the effect of methane content on the induction time is not monotonous. The induction time of the three gas samples is shortened to 0.9–9 min, indicating that NiMnGa particles are suitable for different methane content gases in shortening the induction time. In addition, the effect of NiMnGa particles on gas hydrates with lower methane content is more obvious on shortening the reaction time. However, compared with the influence factors of NiMnGa particle content, methane content is still the primary factor determining the amount of hydrate formation. Therefore, the 3 wt.% NiMnGa system in the G3 gas sample has a strong effect on promoting the kinetic process of gas hydrates.

These findings not only provide a new type of accelerator for the formation of natural gas hydrates; they also open a new way for the application of NiMnGa alloy.

Supplementary Materials: The following supporting information can be downloaded at https://www.mdpi.com/article/10.3390/pr11113149/s1: Table S1: Experimental conditions and results for gas hydrates in different concentrations.

Author Contributions: Investigation, Z.Z. and M.E.; Project administration, B.Z. and Q.W. (Qiang Wu); Resources, Z.L. and H.M.; Supervision, L.L.; Writing—original draft, Z.Z. and X.W.; Writing—review and editing, Q.W. (Qiong Wu), Z.L. and L.L.; Funding acquisition, B.Z. All authors have read and agreed to the published version of the manuscript.

Funding: This research was funded by the National Natural Science Foundation of China (U21A20111, 51804105), the Fundamental Research Funds for the Central Universities (3072022TS1005), and the Fundamental Scientific Research Funds of Provincial Universities in Heilongjiang Province (Hkdcx201902).

Data Availability Statement: The research data is in the Supplementary Information.

Conflicts of Interest: The authors declare no conflict of interest.

References

1. Sloan, E.D., Jr. Fundamental principles and applications of natural gas hydrates. *Nature* **2003**, *426*, 353–359. [CrossRef]
2. Nguyen, N.N.; Nguyen, A.V. Recent insights into the anomalous dual nature (both promotion and Inhibition) of chemical additives on gas hydrate formation. *Chem. Eng. J.* **2023**, *475*, 146362. [CrossRef]
3. Chen, G.J.; Sun, C.Y.; Ma, Q.L. *Gas Hydrate Science and Technology*; Chemical Industry Press: Beijing, China, 2007; pp. 1–30.
4. Lee, Y.; Seo, D.; Lee, S.; Park, Y. Unlocking enhanced gas storage capacity in tuned methane hydrates: Exploring eutectic compositions and water-to-hydrate conversion. *Chem. Eng. J.* **2023**, *475*, 146381. [CrossRef]
5. Abedi-Farizhendi, S.; Iranshahi, M.; Mohammadi, A.; Manteghian, M.; Mohammadi, A.H. Kinetic study of methane hydrate formation in the presence of carbon nanostructures. *Pet. Sci.* **2019**, *16*, 657–668. [CrossRef]
6. Zhang, Y.; Liang, R.D.; Xu, H.J.; Tian, C.S. Recent Advances on experimental study of nucleation process of methane hydrate (Invited). *Acta Photonica Sin.* **2021**, *50*, 82–96.
7. Seo, D.; Lee, S.; Moon, S.; Lee, Y.; Park, Y. Investigating two synthetic routes for gas hydrate formation to control the trapping of methane from natural gas. *Chem. Eng. J.* **2023**, *467*, 143512. [CrossRef]
8. Fan, S.S.; Yu, C.; Lang, X.M.; Wang, Y.H.; Chen, J.B. Micro-Nano-Scale studies on occurrence and gas production and storage technology of marine gas hydrates. *Earth Sci.* **2018**, *43*, 1542–1548.
9. Lang, X.M.; Fan, S.S.; Wang, Y.H.; Li, G.; Yu, C.; Wang, S.L. Opportunities for energy and chemical engineering through clathrate hydrates. *Chem. Ind. Eng. Prog.* **2021**, *40*, 4703–4710.
10. Huang, L.; Tan, J.; Fu, H.; Liu, J.; Chen, X.; Liao, X.; Wang, X.; Wang, C. The non-plane initiation and propagation mechanism of multiple hydraulic fractures in tight reservoirs considering stress shadow effects. *Eng. Fract. Mech.* **2023**, *292*, 109570. [CrossRef]
11. Dontsov, E.V. Analysis of a constant height hydraulic fracture driven by a power-law fluid. *Rock Mech. Bull.* **2022**, *1*, 100003. [CrossRef]
12. Lang, X.M.; Yao, L.M.; Fan, S.S.; Li, G.; Wang, Y.H. Numerical simulation of methane hydrate formation and heat transfer in porous materials. *CIESC J.* **2022**, *73*, 3851–3860.

13. Mirzakimov, U.Z.; Farhadian, A.; Semenov, M.E.; Pavelyev, R.S.; Heydari, A.; Chirkova, Y.F.; Valiullin, L.R. Enhanced methane storage capacity in clathrate hydrate induced by novel biosurfactants: Kinetics, stability, in vivo, and biodegradation investigations. *J. Energy Storage* **2023**, *73*, 108802. [CrossRef]
14. Yan, S.; Huang, J.Y.; Rao, Y.C.; Wang, S.L.; Jia, R.; Liu, B.; Chen, F. Effects of graphene oxide and SDS on CO_2 hydrate formation characteristics. *J. Chang. Univ.* **2020**, *32*, 68–73.
15. Sadeh, E.; Farhadian, A.; Mohammadi, A.; Maddah, M.; Pourfath, M.; Yang, M. Energy-efficient storage of methane and carbon dioxide capture in the form of clathrate hydrates using a novel non-foaming surfactant: An experimental and computational investigation. *Energy Convers. Manag.* **2023**, *293*, 117475. [CrossRef]
16. Qin, Y.; Bao, R.X.; Shang, L.Y.; Zhou, L.; Liu, Z.M. Growth and occurrence characteristics of methane hydrate in a complex system of silica sand and sodium dodecyl sulfate. *Chem. Eng. Sci.* **2021**, *249*, 117349. [CrossRef]
17. Yang, L.; Li, C.; Pei, J.; Wang, X.; Liu, N.; Xie, Y.; Cui, G.; Liu, D. Enhanced clathrate hydrate phase change with open-cell copper foam for efficient methane storage. *Chem. Eng. J.* **2022**, *440*, 135912. [CrossRef]
18. Khandelwal, H.; Fahed, Q.M.; Zheng, J.J.; Venkataraman, P.; Barckholtz, T.A.; Mhadeshwar, A.B.; Linga, P. Effect of L-Tryptophan in promoting the kinetics of carbon dioxide hydrate formation. *Energy Fuels* **2021**, *35*, 649–658. [CrossRef]
19. Liu, X.J.; Ren, J.J.; Chen, D.Y.; Yin, Z.Y. Comparison of SDS and L-Methionine in promoting CO_2 hydrate kinetics: Implication for hydrate-based CO_2 storage. *Chem. Eng. J.* **2022**, *438*, 135504. [CrossRef]
20. Yi, J.; Zhong, D.L.; Yan, J.; Lu, Y.Y. Impacts of the surfactant sulfonated lignin on hydrate based CO_2 capture from a CO_2/CH_4 gas mixture. *Energy* **2019**, *171*, 61–68. [CrossRef]
21. Kastanidis, P.; Tsimpanogiannis, I.N.; Romanos, G.E.; Stubos, A.K.; Economou, I.G. Recent advances in experimental measurements of mixed-gas three-phase hydrate equilibria for gas mixture separation and energy-related applications. *J. Chem. Eng. Data* **2019**, *64*, 4991–5016. [CrossRef]
22. Fang, G.H.; Sun, P.B.; Yu, M.H.; Zhang, W.T.; Tang, X. Research progress and application about paraffin-nanoparticle composite phase-change materials. *Mod. Chem. Ind.* **2022**, *42*, 68–71.
23. Dai, M.L.; Sun, Z.G.; Li, J.; Li, C.M.; Huang, H.F. Progress on promotion technology for gas storage in hydrates. *Chem. Ind. Eng. Prog.* **2020**, *39*, 3975–3986.
24. Liu, G.Q.; Wang, F.; Luo, S.J.; Xu, D.Y.; Guo, R.B. Enhanced methane hydrate formation with SDS-coated Fe_3O_4 nanoparticles as promoters. *J. Mol. Liq.* **2016**, *230*, 315–321. [CrossRef]
25. Ghozatloo, A.; Hosseini, M.; Shariaty-Niassar, M. Improvement and enhancement of natural gas hydrate formation process by Hummers' graphen. *Gas Sci. Eng.* **2015**, *27*, 1229–1233. [CrossRef]
26. Park, S.-S.; Lee, S.-B.; Kim, N.-J. Effect of multi-walled carbon nanotubes on methane hydrate formation. *J. Ind. Eng. Chem.* **2010**, *16*, 551–555. [CrossRef]
27. Arjang, S.; Manteghian, M.; Mohammadi, A. Effect of synthesized silver nanoparticles in promoting methane hydrate formation at 4.7 MPa and 5.7 MPa. *Chem. Eng. Res. Des.* **2013**, *91*, 1050–1054. [CrossRef]
28. Rahmati-Abkenar, M.; Manteghlan, M.; Pahlavazadeh, H. Experimental and theoretical investigation of methane hydrate induction time in the presence of triangular silver nanoparticles. *Chem. Eng. Res. Des.* **2017**, *120*, 325–332. [CrossRef]
29. Rahmati-Abkenar, M.; Manteghlan, M.; Pahlavazadeh, H. Nucleation of ethane hydrate in water containing silver nanoparticles. *Mater. Des.* **2017**, *126*, 190–196. [CrossRef]
30. Aliabadi, M.; Rasoolzadeh, A.; Esmaeilzadeh, F.; Alamdari, A. Experimental study of using CuO nanoparticles as a methane hydrate promoter. *Gas Sci. Eng.* **2015**, *27*, 1518. [CrossRef]
31. Abdi-Khanghah, M.; Adelizadeh, M.; Naserzadeh, Z.; Barati, H. Methane hydrate formation in the presence of ZnO nanoparticle and SDS: Application to transportation and storage. *J. Nat. Gas Sci. Eng.* **2018**, *54*, 120. [CrossRef]
32. Kakati, H.; Mandal, A.; Laik, S. Promoting effect of Al_2O_3/ZnO-based nanofluids stabilized by SDS surfactant on CH_4+C_2H_6+C_3H_8 hydrate formation. *J. Ind. Eng. Chem.* **2016**, *35*, 357–368. [CrossRef]
33. Wang, F.; Song, Y.M.; Liu, G.Q.; Guo, G.; Luo, S.J.; Guo, R.B. Rapid methane hydrate formation promoted by Ag & SDS-coated nanospheres for energy storage. *Appl. Energy* **2018**, *213*, 227–234.
34. Chen, C.; Yuan, H.Y.; Wang, X.M.; Lin, Y.; He, Y.; Wang, F. Recyclable and high-efficiency methane hydrate formation promoter based on SDS-coated superparamagnetic nano-Fe_3O_4. *Chem. Eng. J.* **2022**, *437 Pt 1*, 135365. [CrossRef]
35. Yan, H.C.; Song, X.F.; Xin, F. Storage capacity and duration of methane hydration in a slurry of solid n-Tetradecane. *Energy Fuels* **2014**, *29*, 130–136. [CrossRef]
36. Song, X.F.; XIN, F.; YAN, H.C. Intensification and kinetics of methane hydrate formation under heat removal by phase change of n-tetradecane. *AIChE J.* **2015**, *61*, 3441–3450. [CrossRef]
37. Liu, Y.; Guo, K.; Liang, D.; Fan, S. Effects of magnetic fields on HCFC-141b refrigerant gas hydrate formation. *Sci. China Ser. B Chem.* **2003**, *46*, 407–415. [CrossRef]
38. Sun, S.; Li, Y.; Gu, L.; Yang, Z.; Zhao, J. Experimental study on carbon dioxide hydrate formation in the presence of static magnetic field. *J. Chem. Thermodyn.* **2022**, *170*, 106764. [CrossRef]
39. Yu, L.; Wang, H.Q.; Lu, W.; Sun, H. A lumped parameter model of the longitudinal NiMnGa transducer based on piezomagnetic equations. *J. Acoust. Soc. Am.* **2022**, *152*, 1416–1424.
40. Fan, J.L.; Ai, Y.L.; Ou, Y.S.; Zhu, J.W. Research Progress of Chemical Composition Dependence on the Thermal and Mechanical-Magnetic Properties of NiMnGa-Based Alloys. *Rare Metal Mat. Eng.* **2021**, *50*, 4185–4192.

41. Wu, Q.; Lin, N.Y.; Li, L.; Chen, F.; Xv, X.F.; Zhang, B.Y.; Wu, Q.; Liu, C.H. Synergistic enhancement of coalbed methane hydration process by magnetic field and NiMnGa micro-nano fluid. *Front. Energy Res.* **2022**, *10*, 974647. [CrossRef]
42. Tian, B.; Chen, F.; Liu, Y.; Zheng, Y.F. Structural transition and atomic ordering of $Ni_{49.8}Mn_{28.5}Ga_{21.7}$ ferromagnetic shape memory alloy powders prepared by ball milling. *Mater. Lett.* **2008**, *62*, 2851–2854. [CrossRef]
43. Li, X.S.; Xu, C.G.; Chen, Z.Y.; Wu, H.J. Tetra-n-butyl ammonium bromide semi-clathrate hydrate process for post-combustion capture of carbon dioxide in the presence of dodecyl trimethyl ammonium chloride. *Energy* **2010**, *35*, 3902–3908. [CrossRef]
44. Tian, B.; Chen, F.; Liu, Y.; Zhang, Y.F. Effect of ball milling and post-annealing on magnetic properties of $Ni_{49.8}Mn_{28.5}Ga_{21.7}$ alloy powders. *Intermetallics* **2008**, *16*, 1279–1284. [CrossRef]
45. Tian, B.; Chen, F.; Tong, Y.X. Phase transition of Ni–Mn–Ga alloy powders prepared by vibration ball milling. *J. Alloys Compd.* **2011**, *509*, 4563–4568. [CrossRef]
46. Xu, X.F. Research on Phase Transformation and Magnetic Properties of Nicomnin Magnetic Refrigeration Alloy. Master's Thesis, Harbin Engineering University, Harbin, China, 2020.
47. Tian, B.; Chen, F.; Tong, Y.X. Bending properties of epoxy resin matrix composites filled with Ni-Mn-Ga ferromagnetic shape memory alloy powders. *Mater. Lett.* **2009**, *63*, 1729–1732. [CrossRef]
48. Wu, Q.; Li, Z.; Chen, F.; Lin, N.Y.; Wang, X.Y.; Zhang, B.Y.; Li, L. Ferromagnetic shape memory alloy $Ni_{52.5}Mn_{22.5}Ga_{25}$: Phase transformation at high pressure and effects on natural gas hydration. *Mater. Lett.* **2023**, *343*, 134371. [CrossRef]
49. Pons, J.; Chernenko, V.; Santamarta, R.; Cesari, E. Crystal structure of martensitic phases in Ni-Mn-Ga shape memory alloys. *Acta Mater.* **2000**, *48*, 3027–3038. [CrossRef]

processes

MDPI

Article

Vortex of a Symmetric Jet Structure in a Natural Gas Pipeline via Proper Orthogonal Decomposition

Lihao Li [1,2,3], Jiaxing Lu [1,2,3,*], Haoyu Zhao [1,2,3] and Yilong Qiu [4]

1 Key Laboratory of Fluid and Power Machinery, Ministry of Education, Xihua University, Chengdu 610039, China; leolineo@163.com (L.L.); 17361345933@163.com (H.Z.)
2 School of Energy and Power Engineering, Xihua University, Chengdu 610039, China
3 Key Laboratory of Fluid Machinery and Engineering, Xihua University, Chengdu 610039, China
4 Natural Gas Research Institute, PetroChina Southwest Oil & Gas Field Company, Chengdu 610213, China; qiuyilong@petrochina.com.cn
* Correspondence: jiaxinglu@mail.xhu.edu.cn

Abstract: The impact of particle addition jets on the flow field in natural gas pipelines was investigated, and the structural information of the flow field at different flow velocities in a symmetric jet flow was analyzed via numerical simulation. The results of coherent structures in the high-pressure natural gas pipeline reveal vortex structures of varying sizes both upstream and downstream of the jet flow. To determine the spatial distribution of the main vortex structures in the flow field, proper orthogonal decomposition (POD) mode analysis was performed on the unsteady numerical results. Moreover, the detailed spatial characteristics of the coherent vortex structures represented by each mode were obtained. The results indicate that the large-scale vortex structures within the pipeline are balanced and stable, with their energy increasing as the jet flow velocity increases. Additionally, higher-order modes exhibit significant shedding of small-scale vortex structures downstream of the jet flow. In this research, coherent structures present in symmetric particle addition jets are provided, offering theoretical support for future investigations on the distribution of particle image velocimetry (PIV) flowmeters.

Keywords: jet flow; numerical simulation; POD; natural gas; coherent structure

1. Introduction

Natural gas is an invisible, odorless, and nonrenewable gas mixture, with methane being the major component, accounting for approximately 95% of its composition [1]. Its development and utilization have the potential to generate substantial economic benefits. Natural gas is commonly regarded as a fossil fuel with comparatively low environmental impact, and its use can contribute to environmental quality to some extent. As global concerns about environmental issues continue to grow, the significance of natural gas is anticipated to become increasingly prominent [2–5]. Currently, global natural gas consumption is gradually increasing. The next 15 years are anticipated to witness a continued surge in China's demand for natural gas, and it is estimated that natural gas consumption will reach a peak at approximately 2040, at 6100×10^8 cubic meters [6–8]. As the demand for natural gas on the market continues to grow, it is necessary for the market to optimize the natural gas system and transform natural gas pipeline measurement technology. PIV technology employs the scattering properties of particles in a flow field to track their positions and measure their displacement at different time intervals, thereby enabling the determination of flow velocity and other related parameters [9,10]. PIV technology is playing an increasingly important role in experimental fluid dynamics because of its noninvasive measurement characteristics. Compared with traditional measurement techniques, PIV does not interfere with the flow field, making it advantageous in many fluid experiments [11]. Li et al. [12] argue that PIV technology offers benefits such as resistance to interference and ease of installation. Additionally, they proposed innovative ideas for its application in the measurement of flow fields and developed the PIV flowmeter. However, the distribution of tracer particles in pipeline flow fields is affected by coherent jet vortices, and this principle is not explained in detail. The distribution of particles will affect the accuracy of a PIV flowmeter.

The achievements of computational fluid dynamics (CFD) since the 1970s have demonstrated the powerful vitality of these methods for the study of various fluid phenomena and industrial and engineering applications [13]. Currently, relatively few CFD analyses of symmetric jets exist. Atmaca et al. [14] studied the jets emitted by three different slot nozzles and predicted the free jet velocity at different axial distances. Muhammad et al. [15] used an LES model to study turbulent kinetic energy. Kinetic energy is calculated at the inlet and outlet of turbulent flow. Rehman et al. [16] studied the flow characteristics of a micropolar-Casson fluid, which is doubly stratified and induced by a stretching sheet, their findings are highly consistent with existing results in the literature for specific scenarios. Cao et al. [17] used a new computational method to model the flow of nanofluids in a cavity with a corrugated inner wall with uniform flux. Upadhyay et al. [18] studied the flow and acoustic characteristics of rectangular and elliptical asymmetric jets and compared them with the equivalent area of circular jets. Xiao et al. [19] established a single-nozzle model with different transverse velocities and studied the characteristics of jet flow fields and deflection. Cai et al. [20] developed a two-dimensional computational fluid dynamics (CFDs) model to investigate the flow structure of a liquid nitrogen jet and conducted numerical simulations to analyze the transient velocity and temperature distribution of the jet. Their findings clearly demonstrate that the liquid nitrogen jet exhibits distinct regions, including a velocity boundary and potential core region, which are similar to those observed in water jets. Bi et al. [21] conducted numerical simulations to investigate the characteristics of human cough jets in a calm environment, including the temporal evolution of the velocity distribution, the penetration distance along the jet axis, and the maximum width of the jet. Their findings demonstrate that employing LES methodology enables accurate prediction of the temporal development of a human cough jet. Hou et al. [22] investigated the flow field of an ultrahigh-pressure waterjet (WJ) with high-frequency velocity vibration at the nozzle inlet using computational fluid dynamics. They successfully obtained the velocity field inside the nozzle and examined how various vibration parameters, such as amplitude and frequency, affect the flow field. Their simulation results revealed significant changes in the flow field, corresponding to different frequencies and amplitudes. Lee et al. [23] investigated the three-dimensional density distribution of twin jets through numerical simulations and experimental studies. Their findings revealed that the interaction between the twin jets led to an increase in density within the shear layer located between the nozzles, while both increasing and decreasing interference trends were observed. Furthermore, they noted variations in both the initiation point and growth rate of this interaction phenomenon. Hart et al. [24] performed numerical simulations to investigate the dynamic flow characteristics of oblique-opening rectangular jets and examined the issue of vortex shedding in a cluster of three rectangular burner jets. Their findings indicated that the combination of an adverse pressure gradient, caused by the recess's diffuser-like shape, and fluid entrainment into the gaps between the jets led to the occurrence of vortex shedding. Miltner et al. [25] conducted an extensive examination of the flow characteristics of a turbulent free jet, both through experimental analysis and computational fluid dynamics (CFDs) simulation, using different turbulence models. The findings indicated that numerical simulation proved to be an effective method for precisely characterizing the investigated flow pattern. Zheng et al. [26] investigated the interaction of three jets that were aligned in a single plane and converged towards each other. A particle image velocimetry system was employed to capture the flow fields, facilitating analysis and characterization of the intricate flow patterns on two perpendicular planes based on velocity and vorticity distributions.

CFDs simulations can be conducted for a low cost, have short cycles, and high efficiency. Scholars at home and abroad have used this approach to conduct in-depth research on the internal flow of pipelines [27,28]. Therefore, in this article, unsteady numerical simulations on the symmetric jet flow field of high-pressure natural gas and pipelines have been performed, and the research focuses on investigating the flow properties of natural gas within symmetric jet flow conditions. Among them, the jet has a significant influence on the internal flow field of natural gas, which makes the flow field structure more complex, and it is necessary to reduce the order of the flow field to extract a smooth feature structure from several datasets. Proper orthogonal decomposition (POD) can be employed to effectively reduce the dimensionality of intricate high-dimensional systems [29]. Currently, the POD method is also widely used in the field of computational fluid dynamics. Daichin et al. [30] utilized the POD method to analyze the motion of the near wake behind an elliptic cylinder positioned beneath a free surface. The findings indicated that the first four eigenmodes provide insights into the overall flow structure, with the large-scale structure being related mainly to the most energetic flow motion. Sakai et al. [31] conducted investigations on the velocity fields surrounding a circular cylinder by employing the POD and DMD methods. Both experimentally and numerically, the experimental findings indicated that there was general consistency between the flow structures observed in the POD and DMD modes compared to those obtained through numerical simulations. EI-Adawy et al. [32] employed stereo-PIV to acquire velocity vector fields for in-cylinder flow under various experimental conditions. They utilized the POD method to analyze synthetic velocity vector fields with known characteristics and discovered that, in the case of repeatable flow patterns, only the first mode was adequate for reconstructing the physical properties of the flow. Zhang et al. [33] utilized the large eddy simulation (LES) method to numerically simulate a three-dimensional incompressible flow of an annular jet at a Reynolds number of 8500. The velocity fluctuation vectors were subjected to proper orthogonal decomposition (POD) analysis to investigate the flow dynamics of the wake flow. The findings indicate that the asymmetrical mean flow observed near the annular jet is associated with the first two POD modes, which correspond to a significant displacement in the stagnation point. Jang et al. [34] investigated the flow properties of a turbulent wake formed behind a sphere in a subcritical flow regime. Detailed flow information was obtained using particle image velocimetry measurements and POD modal analysis techniques. Deep et al. [35] investigated the occurrence of vortex shedding in the wake of a circular cylinder and examined how attaching a splitter plate to the rear stagnation point affects this flow phenomenon. Wang et al. [36] employed the LES model to conduct numerical simulations on three typical setups of tornado-like vortices. They also performed POD decomposition on the flow field and observed that the conditional average reveals more significant coherent structures at larger scales compared to the conventional ensemble average. Wu et al. [37] conducted computational fluid dynamics (CFDs) simulations on a horizontal axis wind turbine and analyzed the data using an extended proper orthogonal decomposition (POD) algorithm. The quantitative findings suggest that the wake's tip vortex exhibits a complex spatiotemporal morphological configuration in the higher-order extended POD space, while also confirming the effectiveness and reliability of the obtained radial velocity modes. In addition, Apacoglu et al. [38] conducted a computational analysis on the flow patterns over a two-dimensional circular cylinder and its control through air blowing from various slots positioned on the surface of the cylinder. The POD method was employed to categorize flow structures based on their frequency content. Additionally, an optimal selection of sensor locations on the cylinder surface was determined by employing a one-dimensional POD analysis using surface pressure data obtained from CFD results. Sen et al. [39] conducted a numerical simulation to analyze unsteady two-dimensional laminar fluid flow around a circular cylinder. They qualitatively investigated how changes in the injected stream velocity affected the downstream wake region of the cylinder. Additionally, they utilized proper orthogonal decomposition to identify and quantify dominant modes within the vorticity field, including their corresponding enstrophy contents.

In this article, CFDs is utilized to numerically simulate a high-pressure natural gas pipeline in a multiple-pipe symmetrical injection system. The evolution of vortex structures under different flow velocities of the jet is analyzed, and the study investigates the impact of symmetric injection on the internal flow characteristics of natural gas. The unsteady numerical simulation results are subjected to POD mode analysis, enabling an examination of the flow patterns associated with each mode. A computational approach is employed for conducting the simulations.

2. Numerical Simulation Method

2.1. Numerical Model

In this study, Unigraphics NX was used to create a 1:1 model of the natural gas pipeline, and the outlet of a natural gas pipeline was extended, as shown in Figure 1. The Y+ value of the tube wall is between 30 and 100. The natural gas test system consists of a natural gas main pipeline and four tracer particle injection pipelines, the parameters of which are shown in Table 1. ICEM V14.5 software was used to divide the pipeline model into hexahedral structural grids, and CFX was used to conduct a numerical simulation of the flow field inside the pipeline, as shown in Figure 2. The standard k-epsilon model was used for numerical simulation to simulate the steady-state and transient conditions, and the fluid material was set as methane gas at 1.5 MPa. The initial condition of the transient file is a stable steady-state file, and the unsteady-state flow rate is calculated with a step size of 0.01 s. On this basis, to calculate another 50 steps, the intermediate file was used to calculate the POD. The direction of gravity is shown in Figure 1. The boundary conditions were set as the flow inlet and pressure outlet. The inlet flow rate was set to 100 m^3/h, and the bypass pipe inlet velocities were set to 10 m/s, 20 m/s, 30 m/s, 40 m/s, and 50 m/s.

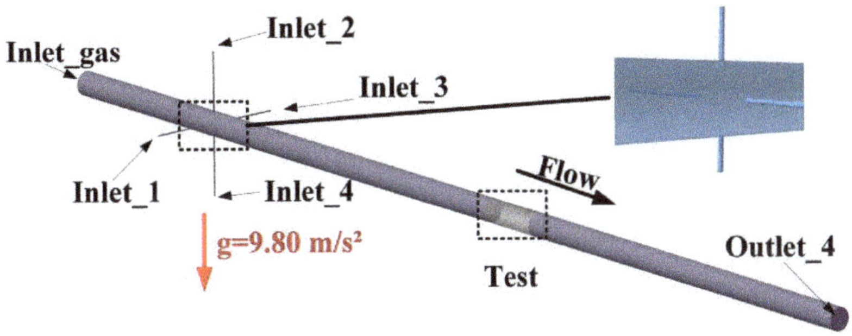

Figure 1. Natural gas pipeline digital model diagram [40].

Table 1. Gas pipeline parameters.

Diameter of Main Pipe (mm)	Main Length (mm)	Length of Bypass Pipe (mm)	Diameter of Bypass Pipe (mm)	Number of Bypass Pipes
100	5000	200	6	4

The standard convergence value is 10^{-6}. The computational fluid shall satisfy the following equation:

Continuity equation:

$$\frac{\partial \rho}{\partial t} + \nabla(\rho \vec{v}) = 0 \tag{1}$$

In this equation, ρ is the fluid density and $\vec{v}$ is the average velocity vector of the mixture.

Momentum equation:

$$\rho \frac{d\vec{v}}{dt} = \nabla \vec{\tau_{ij}} - \nabla p + \rho \vec{F} \tag{2}$$

In this formula, $\vec{\tau_{ij}}$ is the viscous shear stress and $\vec{F}$ is the volumetric force.

Energy equation:

$$\rho\frac{dE}{dt} + p\left(\nabla\vec{v}\right) = \frac{\partial Q}{\partial t} - \nabla\vec{q} + \vec{\Phi} \tag{3}$$

In this equation, Q represents the heat source term and $\vec{\Phi}$ represents the heat dissipation term.

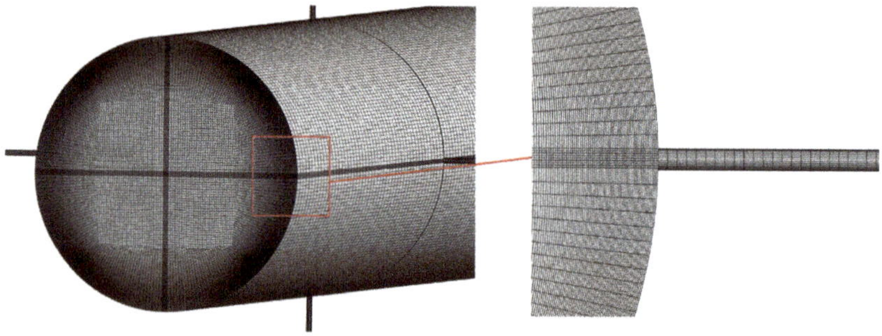

Figure 2. Grid division of bypass pipe in a pipeline.

2.2. *Analysis of the Numerical Result*

In this numerical simulation, a grid-independent study was conducted by monitoring the exit velocity. As depicted in Figure 3a, when the number of grids exceeded 6×10^6, there was no significant change in velocity with an increasing grid count. Since the actual flow rate of the natural gas pipeline is 3.62 m/s, the grid corresponding to the blue scheme was selected for the numerical simulation. This choice ensures an appropriate balance between computational accuracy and efficiency for the given flow conditions. As depicted in Figure 3b, the velocity distribution inside the pipeline measured by experiment is in good agreement with that by numerical simulation.

Figure 4 represents the relative pressure distribution of the jet flow velocity in the range of 10 to 50 m/s in a bypass pipe. When the jet flow velocity is greater than 20 m/s, there is no coherent jet structure in the symmetric jet flow. Figure 4b shows that even if the jet did not collide, the symmetrical pattern of the cross-flow field was disrupted, and the center of the flow velocity began to deviate from the center axis of the main channel. When the flow rate of the jet is higher than 30 m/s, the jet collides violently in the inner center of the pipe, forming a pair of reverse-rotating re-jet vortices upstream. The reason for this is that the mutual collision of the jets results in the maximum relative pressure occurring at the center of the collision, and the airflow velocity near the center approaches a stagnation point. The airflow then spreads out in all directions, forming a vortex structure. The vortexes appear upstream and downstream of the collision center, with the upstream vortex being significantly larger than the downstream vortex. This is because the jet and cross-flow form a larger recirculation zone upstream, and the downstream vortex dissipates more quickly with the cross-flow velocity field. Moreover, the size of the vortex structure formed by the jet and cross-flow affects the downstream turbulence level. The upstream vortex offset also leads to an increase in the downstream turbulence level. Figure 5 shows the distribution of turbulent kinetic energy inside the pipeline, which increases with an increasing jet flow velocity, and the affected range gradually expands.

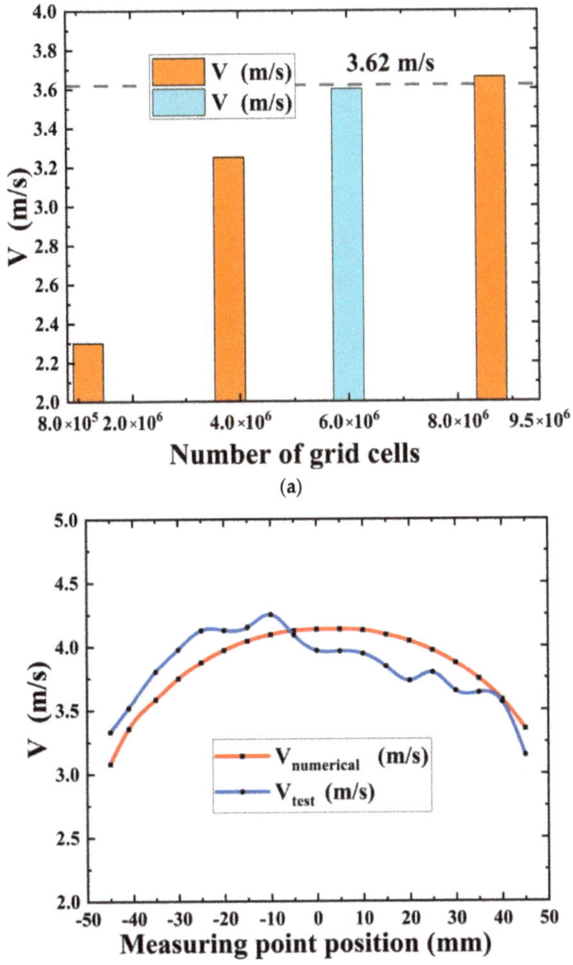

Figure 3. Comparison between the numerical simulation and the experiment. (**a**) Grid independence (the blue one is the selected option). (**b**) Measurement points velocity comparison.

Figure 6 shows the vortex structures under different flow velocities of the jet in a two-dimensional plane based on the Q criterion, including the re-jet vortex, trailing lower vortex, and trailing upper vortex. Figure 7 shows a three-dimensional visualization of the vortex structure. Figure 6 shows that the vorticity increases with an increasing jet flow velocity. The upstream re-jet vortex is the most distinct pair of vortex structures in the flow field. This phenomenon is caused by the mutual collision of jets, which, under the interference and squeezing of the cross-flow, form a re-jet vortex in the upstream part of the jets. With an increasing jet flow velocity, the center position of the vortex structure also migrates upstream. Moreover, even symmetric jet flows can lead to a deviation of the re-jet vortex structure from the center axis of the pipeline. The trailing lower vortex is generated by the jet mixing with the lateral main flow and the mainly low-velocity fluids in the boundary layer near the wall. The trailing upper vortex is generated by the jet wake and migrates downstream with the lateral main flow. When the jet flow velocity is 30 m/s, the vortex structure is located downstream of the jet migrates downward with an increasing cross-flow. However, with an increasing jet flow, the influence of the downstream vortex structure gradually weakens, while the influence range of the vortex structure increases. The upstream re-jet vortex gradually fills the entire flow channel. Notably, the turbulent kinetic energy and vortex structure exhibit opposite characteristics, because the turbulent region has a higher flow velocity, while a vortex structure is formed due to the rotation of the velocity field. In the flow field of the pipeline, symmetric jets collide with each other, causing the radial velocity of the jets to cancel. There is a strong shear force between the jets and the cross-flow, leading to an increase in the vorticity of the velocity field.

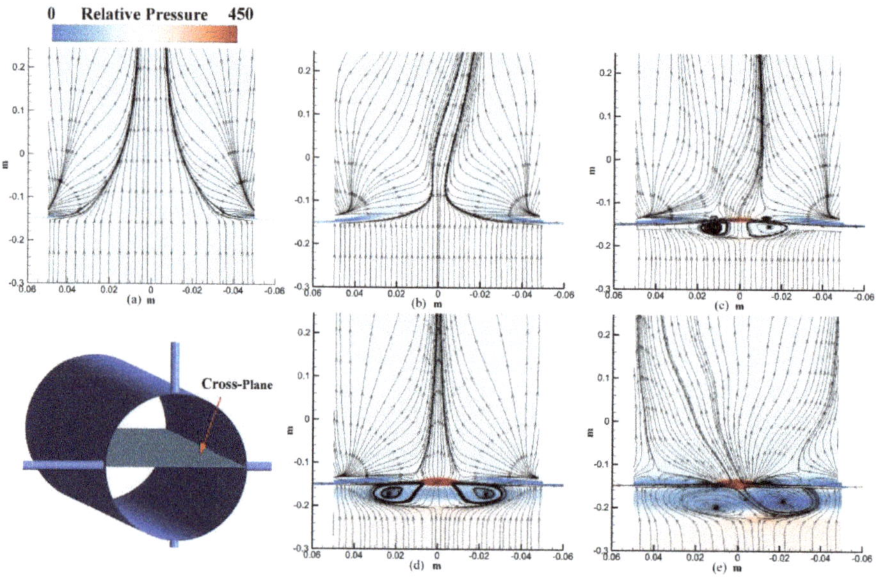

Figure 4. Relative pressure distribution at 10 m/s (**a**), 20 m/s (**b**), 30 m/s (**c**), 40 m/s (**d**), and 50 m/s (**e**).

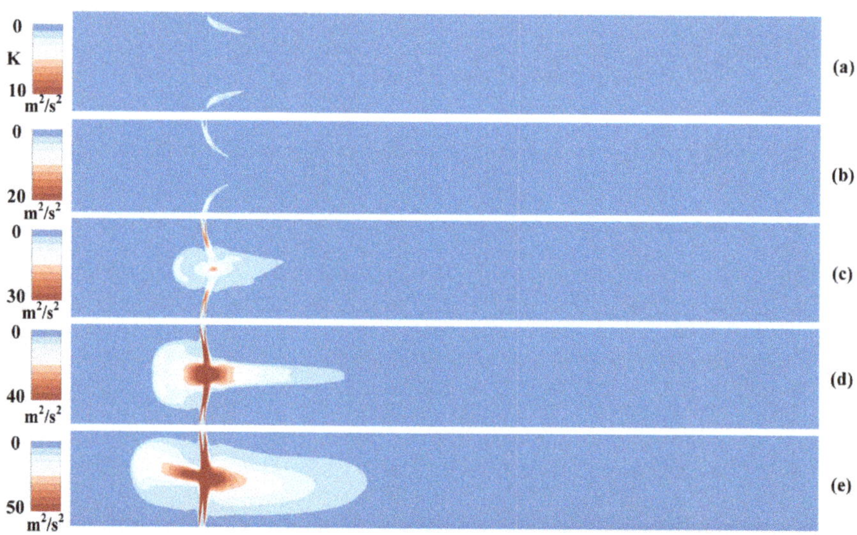

Figure 5. The distribution of turbulent kinetic energy at 10 m/s (**a**), 20 m/s (**b**), 30 m/s (**c**), 40 m/s (**d**), and 50 m/s (**e**).

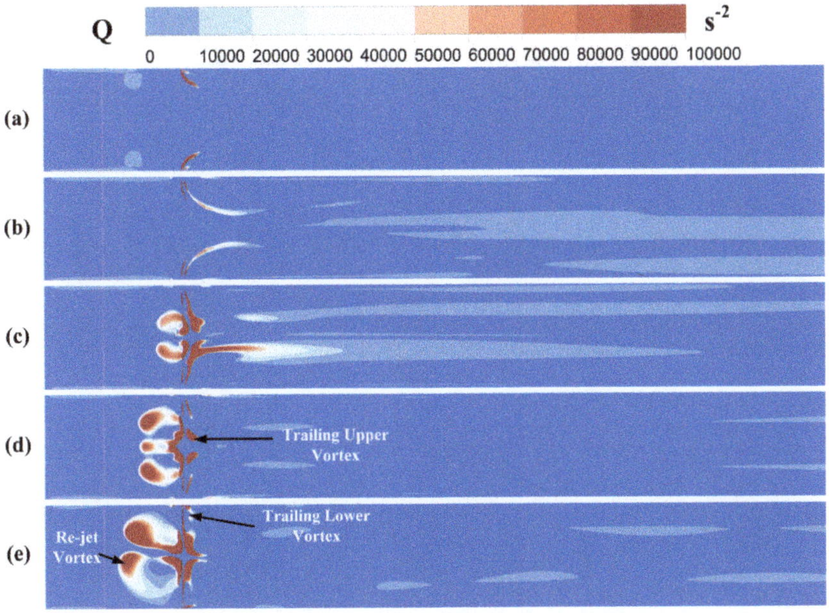

Figure 6. The distribution of the vortex at 10 m/s (**a**), 20 m/s (**b**), 30 m/s (**c**), 40 m/s (**d**), and 50 m/s (**e**).

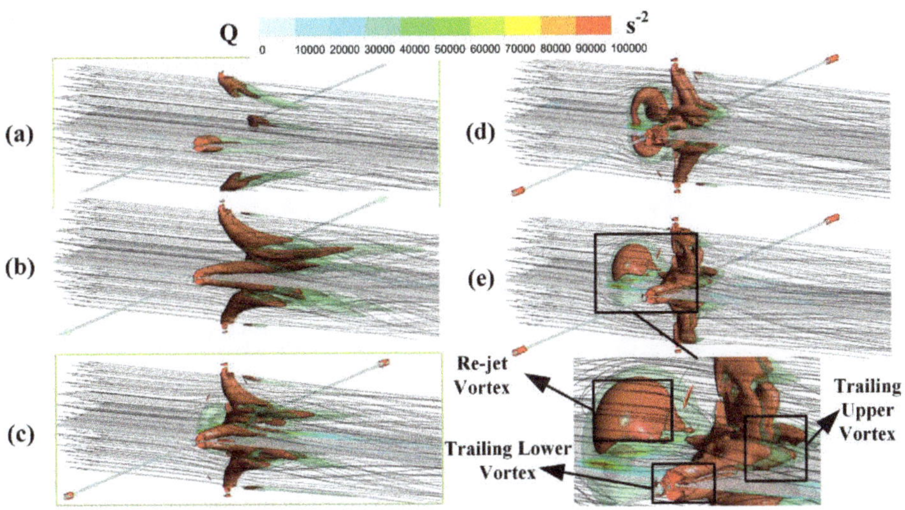

Figure 7. The distribution of the 3D vortex at 10 m/s (**a**), 20 m/s (**b**), 30 m/s (**c**), 40 m/s (**d**), and 50 m/s (**e**).

3. Proper Orthogonal Decomposition

To better analyze and extract data from the symmetric jet flow field, proper orthogonal decomposition is used in this article. The POD method can be used to obtain low-dimensional characteristics of flow fields from large amounts of high-dimensional flow data. Considering the distinct flow characteristics of symmetric jet collisions, three pressure monitoring points are established inside the natural gas pipeline. Fifty snapshots containing flow field features (one transient file every 0.01 s, totaling 0.5 s), were calculated, and extracted, and POD flow field information extraction was performed. Considering the complex structure and magnanimous space nodes, the snapshot POD method is adopted for its advantages in unsteady flow field analysis. This method is suitable for complex problems, where the number of spatial points M is much greater than the number of sampling time points. The mathematical derivation process is introduced below.

It is supposed that there is a set of observation data matrices that vary with time as follows:

$$X = [x_1 \quad x_2 \quad \cdots x_n] \in R^{m \times n} \tag{4}$$

where m represents the size of the spatial domain, and can be used to establish the spatial correlation matrix with Δt:

$$S_{m\times n} = \frac{1}{n} X_{m\times n} X_{m\times n}^T \tag{5}$$

$S_{m\times n}$ is a real symmetric matrix of order $m \times n$, the eigenvalues of which are greater than or equal to 0, with the eigenvectors corresponding to different eigenvalues. The eigenvalues λ and eigenvectors ϕ are solved as follows:

$$S\phi = \lambda\phi \tag{6}$$

The resulting feature vector ϕ_j is the J-order POD mode, and the corresponding time coefficient is:

$$(a_j(t))^T = (X(x,t))^T \phi_j(x) \tag{7}$$

The $r(r > 0)$ nonzero eigenvalues can be further rewritten as follows:

$$\lambda_j = \phi_j^T S(\varphi_j) = \phi_j^T X X^T(\varphi_j) = \left| X^T \varphi_j \right|^2 \tag{8}$$

The eigenvalue is the square of the modulus projected by the original function onto the eigenmode, and its value represents the energy ratio of the mode. The corresponding energy content of each order mode ratio can be expressed as follows:

$$\varepsilon_j = \frac{\lambda_j}{\sum\limits_{j=1}^{n} \lambda_j} \tag{9}$$

3.1. Vortex Distribution Characteristics

In this article, the POD method is used to analyze coherent vortex structures in a jet flow field. Figure 8 shows the energy proportions of modes of order 1–10 at different jet flow rates. At flow rates of 50 m/s and 40 m/s, the first- and second-order modes account for 93.7% and 95.7%, respectively, of the total energy, which is related to the formation of large vortex structures by jet impact. At a 30 m/s jet flow rate, the first- and second-order modes account for 99% of the total energy, and the first-mode energy does not account for more than 85%, as at 50 m/s and 40 m/s, mainly because the symmetrical jets at 30 m/s interfere with each other to form a re-jet vortex. At a flow rate of 20 m/s, the proportion of the mode energy reaches 92.7%. Although the jets do not collide, they interfere with the flow field. At a flow rate of 20 m/s, the energy of each order decreases step by step, and there is no dominant mode. The low-order POD mode represents the large-scale coherent vortex structure of the flow field in the pipeline, and the high-order POD mode represents the small-scale vortex structure of the flow field in the pipeline.

Figure 9a shows the first six orders of the flow velocity at 10 m/s. In Figure 8, the proportion of first-mode energy does not reach 80% or more, as in other percentages. The trailing lower and upper vortexes are the main components of the vortex. The first-order mode mainly manifests as the trailing upper vortex, which exhibits incomplete symmetry. The trailing lower vortex has relatively weak energy, and its detachment has a significant impact on the boundary layer in the downstream flow path. The lower vortex of the fifth-order mode clearly exhibits regular shedding. Figure 9b shows the first six modes of the flow velocity at 20 m/s. The jet flow has not yet collided, and the main vortex is mainly concentrated in the trailing upper vortex, which is highly concentrated at the tail of the jet flow. Notably, the trailing upper vortex can affect the formation of wall-mounted vortex structures in the boundary layer at certain positions downstream of the nozzle and continue to influence subsequent flow fields.

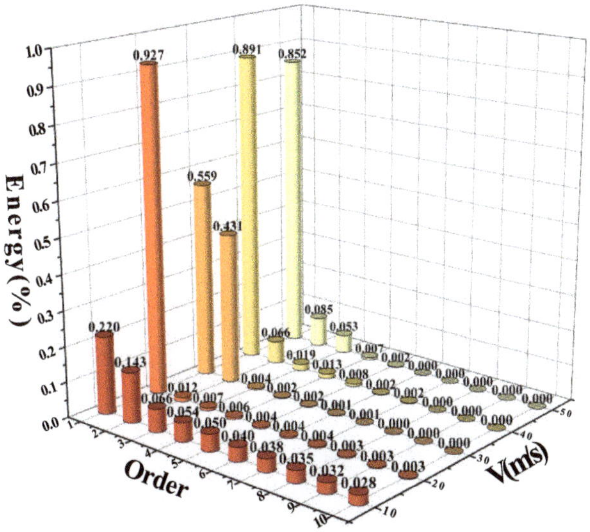

Figure 8. The energy proportion of modes 1–10 at different jet flow velocities.

Figure 10a shows the first six modes at a jet flow rate of 30 m/s. As shown in the figure, the jets collide on the central axis of the pipeline at this time, and the first-order modal diagram shows that an obvious re-jet vortex is formed upstream of the collision center point. Downstream of the nozzle, trailing upper and lower vortexes also form. Based on the energy proportion graph, the energy ratio of the re-jet vortex and trailing upper vortex under the first-order mode reaches 99%. Moreover, many small vortex structures upstream and downstream can be captured on the higher-order modal diagram. Figure 10b shows the first six modes at a jet flow rate of 40 m/s. The first- and second-order modes show the upstream region of the jet hedge center, and the large vortex structure on both sides of the pipeline central axis is consistent with the re-jet vortex in Figure 4e. The first-order mode can clearly capture the existence of trailing upper and lower vortexes, and the energy of the upstream re-jet vortex is concentrated in the jet hedge center. Figure 10c shows the vortex structure of the first six modes at a jet flow speed of 50 m/s. The vortex structure is analyzed by the energy distribution and the structural modal diagram shown. With an increasing jet velocity, the upstream re-jet vortex structure gradually expands, and the downstream vortex structure becomes more concentrated. In the first six modes, the upstream region of the jet hedge center and both sides of the pipeline central axis exhibit a large re-jet vortex structure, which is the main vortex structure upstream. The modes are symmetrical along the central axis, which represents the main vortex structure in the flow field under these conditions. Figure 11 shows the time domain plot of the first three POD mode coefficients. There is a 90-degree phase difference between the first and second modes.

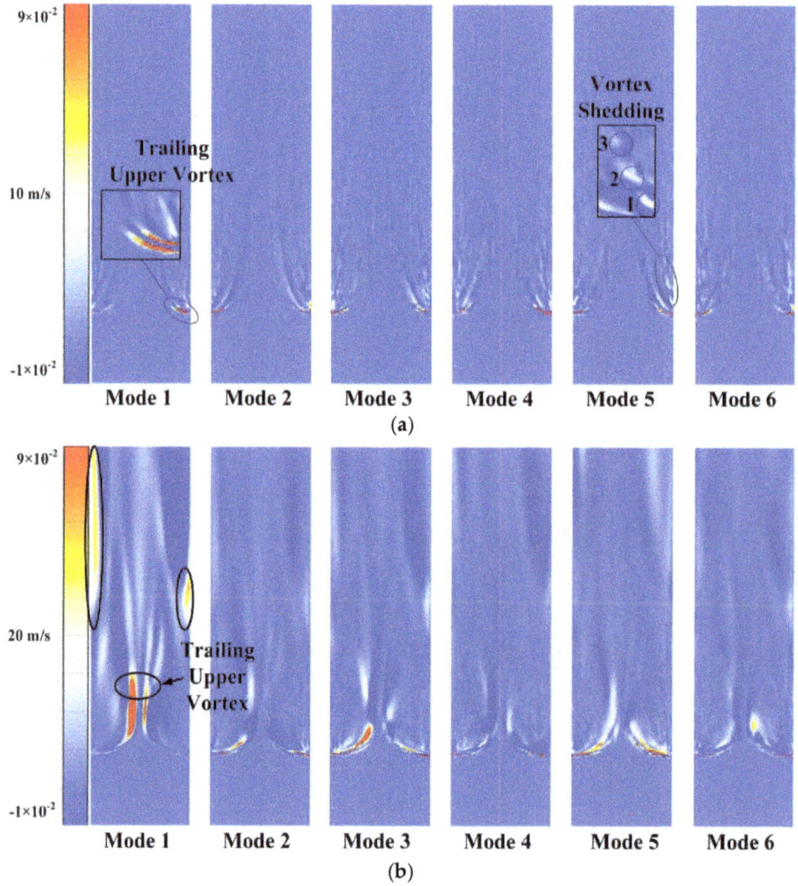

Figure 9. The first six orders at flow rates of 10 m/s (**a**) and 20 m/s (**b**).

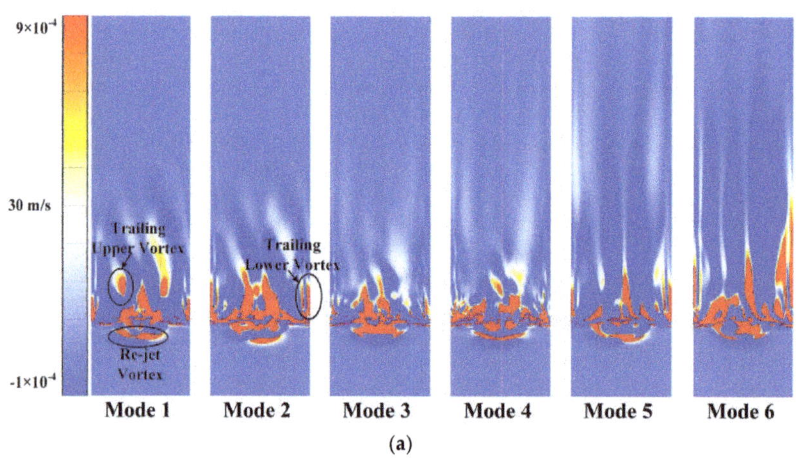

Figure 10. *Cont.*

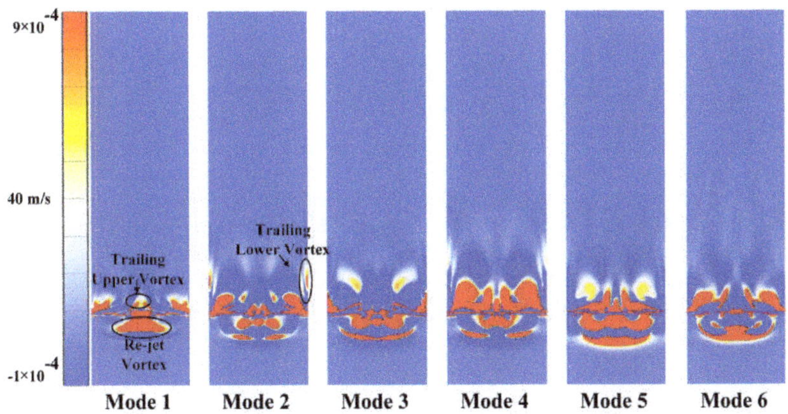

(b)

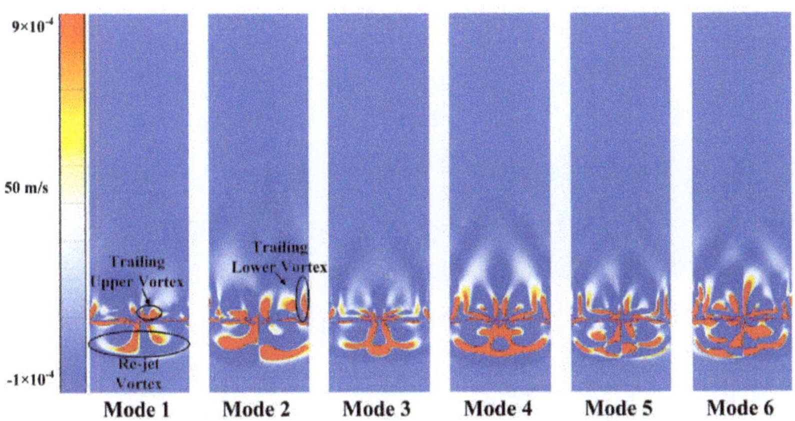

(c)

Figure 10. The first six orders at a flow rate of 30 m/s (**a**) 40 m/s (**b**) and 50 m/s (**c**).

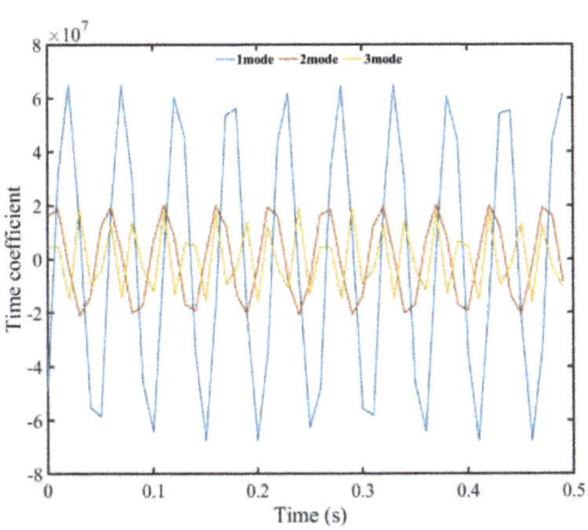

Figure 11. The first six orders at a flow rate of 50 m/s.

3.2. The Result of PIV

Figure 12 is a photograph of the joint test between the Key Laboratory of Fluid and Power Machinery of the Ministry of Education of Xihua University and the Chengdu Branch of the National Oil and Natural Gas Large Flow Metering Station. Figure 13a shows the distribution of tracer particles caused by the injection jet, where the tracer particles are evenly mixed in the pipeline. Figure 13b shows the PIV measurement velocity flow field, which has a uniform velocity distribution and accurate measurement. Therefore, the vortex structure caused by the symmetrical jet filling mode is conducive to ensuring the distribution of tracer particles and improving the snapshot quality of the PIV flowmeter.

Figure 12. The PIV flowmeter for natural gas applications.

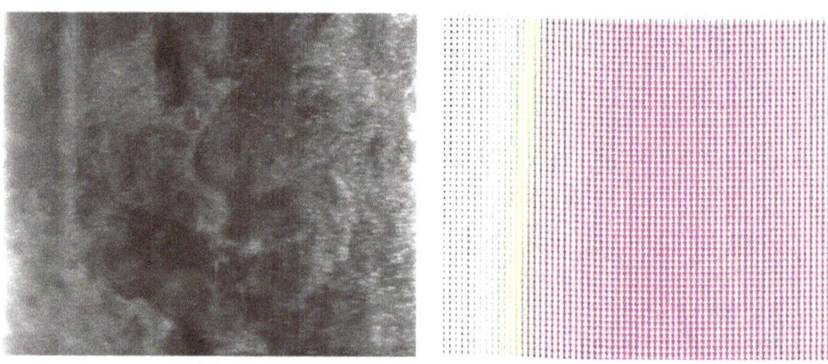

(a) (b)

Figure 13. The test of PIV: (**a**) The distribution of tracer particles caused by the injection jet. (**b**) The PIV measurement velocity flow field.

4. Conclusions

The effect of a jet in a pipeline is completely different to that in typical free flow. Gas flowing through the pipeline is predominantly methane, and the flow environment is characterized by high pressure. The flow field within the pipeline is constrained, meaning that both the main flow and the jet streams develop within a limited empty space. As a result, there exists a coherent structure between the jet and the cross flow, which influences the dynamics and behavior of the flow within the pipeline. The main conclusions are as follows:

1. The jet flow interacts with the crossflow, resulting in the formation of trailing upper and lower vortices. At lower jet flow rates, the trailing upper and lower vortices are located above and below the nozzle, respectively.
2. As the velocity of the jet flow increases, the trailing upper vortices begin to interfere with each other and gradually merge, forming a re-jet vortex. This observation highlights the complex dynamics and interactions between the jet flow and the surrounding crossflow within the pipeline.
3. The POD method is employed to analyze the coherent vortex structure in different modes. The first and second order modes exhibit relatively high energy, enabling them to accurately capture the coherent vortex structure of the jet.
4. Higher order modes are useful for observing and studying the growth and shedding of small-scale vortex structures.
5. The vortex structure formed by symmetrical jet is conducive to PIV flowmeter measurements.

Author Contributions: Conceptualization, L.L.; writing—original draft preparation, J.L., H.Z. and L.L.; writing—review and editing, Y.Q. All authors have read and agreed to the published version of the manuscript.

Funding: This work was funded by the National Natural Science Foundation of China (No. 52009115 and No. U23A20669).

Data Availability Statement: The data presented in this study are available in the main text of the article.

Conflicts of Interest: Yilong Qiu was employed by PetroChina Southwest Oil & Gas Field Company. The remaining authors declare that the research was conducted in the absence of any commercial or financial relationships that could be construed as a potential conflict of interest.

References

1. Neumann, A.; Hirschhausen, C.V. Natural Gas: An Overview of a Lower-Carbon Transformation Fuel. *Rev. Environ. Econ. Policy* **2015**, *9*, 64–84. [CrossRef]
2. Wu, X.; Wang, Z.; Dong, M.; Dong, L.F.; Ge, Q. A Critical Analysis of Natural Gas Liquefaction Technology. *Fluids Dyn. Mater. Process.* **2022**, *18*, 145–158. [CrossRef]
3. Kuang, Y.; Lin, B. Unwatched Pollution Reduction: The Effect of Natural Gas Utilization On Air Quality. *Energy* **2023**, *273*, 127247. [CrossRef]
4. Kuang, Y.M.; Lin, B.Q. Natural Gas Resource Utilization, Environmental Policy and Green Economic Development: Empirical Evidence From China. *Resour. Policy* **2022**, *79*, 102992. [CrossRef]
5. Li, W.; Lu, C. The Multiple Effectiveness of State Natural Gas Consumption Constraint Policies for Achieving Sustainable Development Targets in China. *Appl. Energy* **2019**, *235*, 685–698. [CrossRef]
6. Jia, A.; Cheng, G.; Chen, W.; Li, Y. Forecast of natural gas supply and demand in China under the background of "Dual Carbon Targets". *Pet. Explor. Dev.* **2023**, *50*, 492–504. [CrossRef]
7. Liu, H.; Liu, Y.; Wang, C.; Song, Y.; Jiang, W.; Li, C.; Zhang, S.; Hong, B. Natural Gas Demand Forecasting Model Based On Lasso and Polynomial Models and its Application: A Case Study of China. *Energies* **2023**, *16*, 4268. [CrossRef]
8. Ou, J.; Chen, Y. A Novel Dynamic Parameter Discrete Grey Model and its Application. *Energy Rep.* **2023**, *9*, 4941–4950. [CrossRef]
9. Jenkins, C.M.; Horie, Y.; Wu, C.Y. Particle Velocity and Structures in Blast Waves Imaged Using Particle Image Velocimetry. *Int. J. Multiph. Flow* **2010**, *36*, 88–91. [CrossRef]
10. Gao, Q.; Wang, H.; Shen, G. Review On Development of Volumetric Particle Image Velocimetry. *Chin. Sci. Bull.* **2013**, *58*, 4541–4556. [CrossRef]
11. Liu, S.; Xu, J.; Gong, H.; Yu, K. Experimental Comparison of Piv-Based Pressure Measurements in Supersonic Flows with Shock Waves. *Meas. Sci. Technol.* **2019**, *30*, 105201. [CrossRef]
12. Li, W.; Li, Z.; Deng, W.; Ji, L.; Qiu, Y.; Chen, H. Particle Image Velocimetry Flowmeter for Natural Gas Applications. *Flow Meas. Instrum.* **2021**, *82*, 102072.
13. Ranz, W.E. Some Experiments On Orifice Sprays. *Can. J. Chem. Eng.* **1958**, *36*, 175–181. [CrossRef]
14. Atmaca, M.; Cetin, B.; Ezgi, C.; Kosa, E. CFD Analysis of Jet Flows Ejected From Different Nozzles. *Int. J. Low-Carbon Technol.* **2021**, *16*, 940–945. [CrossRef]
15. Muhammad, N.; Alharbi, K.A.M. OpenFOAM for Computational Hydrodynamics Using Finite Volume Method. *Int. J. Mod. Phys. B* **2023**, *37*, 2350026. [CrossRef]

16. Rehman, S.; Muhammad, N. Mathematical Analysis of Nonlinear Models and their Role in Dynamics. *Mod. Phys. Lett. B* **2023**. [CrossRef]
17. Cao, W.; Altoum, S.H.; Othman, H.A.; Alzubaidi, A.M.; Alghawli, A.S. Numerical Modeling for Thermal Behavior of Nanomaterial Laminar Flow and Convective Heat Transfer in Appearance of Magnetic Field. *Case Stud. Therm. Eng.* **2023**, *42*, 102727. [CrossRef]
18. Upadhyay, P.; Valentich, G.; Kumar, R.; Alvi, F. Flow and Acoustic Characteristics of Non-Axisymmetric Jets at Subsonic Conditions. *Exp. Fluids* **2017**, *58*, 52. [CrossRef]
19. Xiao, S.; Ge, Z.; Ren, Q.; Liu, J.; Wang, H. Numerical Analysis On the Flow Field Structure and Deflection Characteristics of Water Jets Under Nozzle Moving Conditions. *Eng. Appl. Comp. Fluids* **2020**, *14*, 1279–1301. [CrossRef]
20. Cai, C.; Ren, K.; Li, Q. Numerical Simulation of the Flow Field Structure of Liquid Nitrogen Jet. *Therm. Sci.* **2019**, *23*, 1337–1343. [CrossRef]
21. Bi, R.; Ali, S.; Savory, E.; Zhang, C. A Numerical Modelling Investigation of the Development of a Human Cough Jet. *Eng. Comput.* **2022**, *39*, 773–791. [CrossRef]
22. Hou, R.; Huang, C.; Zhu, H. Numerical Simulation Ultrahigh Waterjet (Wj) Flow Field with the High-Frequency Velocity Vibration at the Nozzle Inlet. *Int. J. Adv. Manuf. Tech.* **2014**, *71*, 1087–1092. [CrossRef]
23. Lee, C.; Ozawa, Y.; Haga, T.; Nonomura, T.; Asai, K. Comparison of Three-Dimensional Density Distribution of Numerical and Experimental Analysis for Twin Jets. *J. Vis.* **2021**, *24*, 1173–1188. [CrossRef]
24. Hart, J.T.; Karim, M.R.; Bhuiyan, A.A.; Witt, P.; Naser, J. Numerical Modelling of Unsteady Flow Behaviour in the Rectangular Jets with Oblique Opening. *Alex. Eng. J.* **2016**, *55*, 2309–2320. [CrossRef]
25. Miltner, M.; Jordan, C.; Harasek, M. CFD Simulation of Straight and Slightly Swirling Turbulent Free Jets Using Different Rans-Turbulence Models. *Appl. Therm. Eng.* **2015**, *89*, 1117–1126. [CrossRef]
26. Zheng, J.X.; Wang, X.K.; Tan, D.S.; Tan, S.K. Experimental Investigation of Interactions of Three Co-Planar and Converging Jets. *Exp. Fluids* **2014**, *55*, 1733. [CrossRef]
27. Afzal, A.; Ansari, Z.; Faizabadi, A.R.; Ramis, M.K. Parallelization Strategies for Computational Fluid Dynamics Software: State of the Art Review. *Arch. Comput. Methods Eng.* **2017**, *24*, 337–363. [CrossRef]
28. Dai, Y.; Zhang, Y.; Li, X. Numerical and Experimental Investigations On Pipeline Internal Solid-Liquid Mixed Fluid for Deep Ocean Mining. *Ocean Eng.* **2021**, *220*, 108411. [CrossRef]
29. Kang, H.; Tian, Z.; Chen, G.; Li, L.; Wang, T. Application of Pod Reduced-Order Algorithm On Data-Driven Modeling of Rod Bundle. *Nucl. Eng. Technol.* **2022**, *54*, 36–48. [CrossRef]
30. Lee, S.J. Pod Analysis of Near-Wake Structures of an Elliptic Cylinder Adjacent to a Free Surface. *J. Vis.* **2004**, *7*, 179–186.
31. Sakai, M.; Sunada, Y.; Imamura, T.; Rinoie, K. Experimental and Numerical Studies on Flow behind a Circular Cylinder Based on POD and DMD. *Trans. Jpn. Soc. Aeronaut. Space Sci.* **2015**, *58*, 100–107. [CrossRef]
32. El-Adawy, M.; Heikal, M.R.; Aziz, A.R.A.; Adam, I.K.; Ismael, M.A.; Babiker, M.E.; Baharom, M.B.; Firmansyah; Abidin, E.Z.Z. On the Application of Proper Orthogonal Decomposition (Pod) for in-Cylinder Flow Analysis. *Energies* **2018**, *11*, 2261. [CrossRef]
33. Zhang, Y.; Vanierschot, M. Proper Orthogonal Decomposition Analysis of Coherent Motions in a Turbulent Annular Jet. *Appl. Math. Mech.* **2021**, *42*, 1297–1310. [CrossRef]
34. Jang, Y.I.; Lee, S.J. Pod Analysis On the Sphere Wake at a Subcritical Reynolds Number. *J. Vis.* **2010**, *13*, 311–318. [CrossRef]
35. Deep, D.; Sahasranaman, A.; Senthilkumar, S. POD Analysis of the Wake Behind a Circular Cylinder with Splitter Plate. *Eur. J. Mech. B Fluids* **2022**, *93*, 1–12. [CrossRef]
36. Wang, M.; Cao, S.; Cao, J. POD-Based Analysis of Time-Resolved Tornado-Like Vortices. *Wind Struct.* **2021**, *33*, 13–27.
37. Wu, W.; Zhou, C. A Numerical Study of the Tip Wake of a Wind Turbine Impeller Using Extended Proper Orthogonal Decomposition. *Fluid Dyn. Mater. Process.* **2020**, *16*, 883–901. [CrossRef]
38. Apacoglu, B.; Paksoy, A.; Aradag, S. CFD Analysis and Reduced Order Modeling of Uncontrolled and Controlled Laminar Flow Over a Circular Cylinder. *Eng. Appl. Comput. Fluid Mech.* **2011**, *5*, 67–82. [CrossRef]
39. Sen, U.; Mukhopadhyay, A.; Sen, S. Effects of Fluid Injection on Dynamics of Flow Past a Circular Cylinder. *Eur. J. Mech. B Fluids* **2017**, *61*, 187–199. [CrossRef]
40. Chen, H.; Qiu, Y.; Wang, H. Numerical and Experimental Investigation of Flow Characteristics in Natural Gas Pipe. *AIP Adv.* **2023**, *13*, 045017. [CrossRef]

Article

Adsorption and Diffusion Properties of Gas in Nanopores of Kerogen: Insights from Grand Canonical Monte Carlo and Molecular Dynamics Simulations

Shouheng Xiao [1], Xiugang Liu [2], Yun Li [3,*], Qiang Zheng [1], Ning Wang [1], Yun Qiao [4], Youyin Zhang [1] and Chuanjun Yi [1]

1 Research Institute of Exploration & Development, Xinjing Oilfield Company, PetroChina, Karamay 834000, China
2 School of Emergency Management and Safety Engineering, China University of Mining and Technology, Beijing 100083, China
3 Academy for Advanced Interdisciplinary Studies, Southern University of Science and Technology, Shenzhen 518055, China
4 Research Institute of Engineering Technology, Xinjing Oilfield Company, PetroChina, Karamay 834000, China
* Correspondence: liy39@sustech.edu.cn

Abstract: Investigating the adsorption and diffusion processes of shale gas within the nanopores of kerogen is essential for comprehending the presence of shale gas in organic matter of shale. In this study, an organic nanoporous structure was constructed based on the unit structure of Longmaxi shale kerogen. Grand canonical Monte Carlo and molecular dynamics simulation methods were employed to explore the adsorption and diffusion mechanisms of pure CH_4, CO_2, and N_2, as well as their binary mixtures with varying mole fractions. The results revealed that the physical adsorption characteristics of CH_4, CO_2, and N_2 gases on kerogen adhered to the Langmuir adsorption law. The quantity of adsorbed gas molecules increased with rising pressure but decreased with increasing temperature. The variation in the heat of adsorption was also analyzed. Under identical temperature and pressure conditions, the adsorption of CH_4 increased with higher mole fractions of CH_4, whereas it decreased with greater mole fractions of CO_2 and N_2. Notably, CO_2 molecules exhibited a robust interaction with kerogen molecules compared to the adsorption properties of CH_4 and N_2. Furthermore, the self-diffusion coefficient of gas within kerogen nanopores gradually decreased with increasing pressure or decreasing temperature. The diffusion capacity of gas molecules followed the descending order $N_2 > CH_4 > CO_2$ under the same pressure and temperature conditions.

Keywords: adsorption and diffusion; grand canonical Monte Carlo; molecular dynamics simulation; kerogen; shale gas

1. Introduction

With the ongoing surge in global energy consumption, there has been a heightened focus on the exploration and development of oil and natural gas in recent years. Shale gas, an unconventional resource, has emerged as a focal point in the global quest for natural gas resources. Its exploration and utilization are pivotal for future energy needs [1,2]. Simultaneously, shale gas plays a crucial role in the overall landscape of oil and natural gas development and is a cornerstone in national strategic energy reserves, garnering significant attention worldwide [3,4].

Shale gas refers to the natural gas stored in shale formations, characterized by its richness in organic matter (e.g., kerogen) and clay minerals (e.g., montmorillonite, illite, kaolinite, chlorite, etc.). While a small portion of gas dissolves in the formation water, the majority of free gas is present in large pores and natural cracks, with a significant amount of adsorbed gas stored in the nanopores of clay mineral and kerogen particles [5–7]. The development of nanopores in shale reservoirs results in an exceptionally large specific surface area for shale. Consequently, the adsorption capacity of shale gas can range from 20% to 80% [8,9]. Understanding the diffusion process of desorbed gas in pores is crucial for shale gas migration as pressure decreases. The complexity of shale gas diffusion arises from the presence of numerous nanopores formed by clay minerals and organic matter. Consequently, the adsorption properties of shale gas are influenced not only by the physical and chemical properties of clay minerals but also by external conditions such as pressure, temperature, and water content [7,10–12]. Previous research on shale gas adsorption characteristics has predominantly relied on isotherm adsorption experiments [11,13] and isotherm adsorption models [14–16]. Previous studies have developed new adsorption models based on extensive experimental data, offering more realistic representations of shale gas adsorption in actual scenarios [17,18]. However, due to the limitations of experimental tests, the variation in adsorption amounts has not been fully explained at the molecular level, particularly concerning kerogen nanopores. The diffusion characteristics of shale gas have primarily been investigated through experimental tests of diffusion coefficients, revealing variations under different temperature and pressure conditions [19,20]. Mathematical models based on Fick's law have been established to describe shale gas diffusion, considering the effects of reservoir conditions, gas types, and other factors [21,22]. Despite analyzing the diffusion properties of shale gas in nanopores through mathematical models, these approaches fall short of providing a comprehensive micro-scale explanation for the variation in diffusion coefficients and diffusion flux of shale gas. Accurately revealing the adsorption patterns and mechanisms of shale gas in the nanopores of shale from a molecular level remains challenging. With the rapid development of computer technology, molecular simulation methods have become a powerful tool for studying the properties of fluids in confined spaces. These methods can provide insights into adsorption structures and dynamic processes at the molecular level, which are difficult to observe and detect experimentally. Grand canonical Monte Carlo (GCMC) and molecular dynamics (MD) simulations have been used to provide insights into the mineral and kerogen structures, microscopic adsorption [23–25], diffusion [26,27], flow [28], and transport properties of fluids [29–31]. For example, Julien Collenll et al. [32] established the structure of type II kerogen based on the proportions of various elements. The GCMC and MD methods were used to study the pore size distribution and diffusion properties of fluids in kerogen systems. Philippe Ungerer et al. [33] employed molecular simulation methods to create the molecular structure of type I, II, and III kerogen based on their C/H and C/O ratios, investigating the thermodynamics properties and maturities of various types of kerogen. Such simulations offer a reliable foundation for understanding the interaction between kerogen and gas fluids in natural gas reservoirs. The Longmaxi formation in the Sichuan Basin serves as the primary shale exploration and development block in China. While numerous studies have investigated the physical and geochemical characteristics as well as the pore evolution of shale samples in this region over the past few years, few studies have delved into the microscopic mechanisms of shale gas adsorption and diffusion within the organic matter of Longmaxi shale.

In this study, the adsorption and diffusion properties of shale gas in the kerogen nanopores of Longmaxi shale were investigated using GCMC and MD simulations. A three-dimensional model comprising five kerogen molecules was constructed for the analysis. The investigation focused on the adsorption and diffusion of pure CH_4, CO_2, and N_2, along with their binary mixtures at various mole fractions. Calculations were conducted to determine the adsorption amount, isosteric heat, self-diffusion coefficient, and radial distribution function. The microscopic adsorption and diffusion properties of these gases in kerogen were thoroughly analyzed.

2. Simulation Method

2.1. Kerogen Model

Utilizing the average molecular structure as a reference [34], the simulation model for shale gas adsorption and diffusion was constructed based on the Longmaxi kerogen molecule (Type II kerogen: O/C = 0.092; H/C = 0.77) [35]. This molecular structure predominantly consists of aromatic hydrocarbons, including benzene rings, hydroxyl groups, methylene, methyl, hydroxyl groups, and various other functional groups. Alkene and carbonyl groups serve as the structural units linking each aromatic hydrocarbon. It has been found that kerogen functional groups containing nitrogen (N) and oxygen (O) elements increase the CH_4 adsorption capacity, while functional groups containing methyl groups have a weak impact on CH_4 adsorption capacity [36]. In this study, the chemical formula of kerogen is $C_{206}H_{158}O_{19}N_4S_4$. The kerogen molecular structure was generated, and the initial model underwent annealing optimization from a high temperature to a low temperature to achieve the lowest energy configuration. A snapshot of the kerogen structure is depicted in Figure 1a.

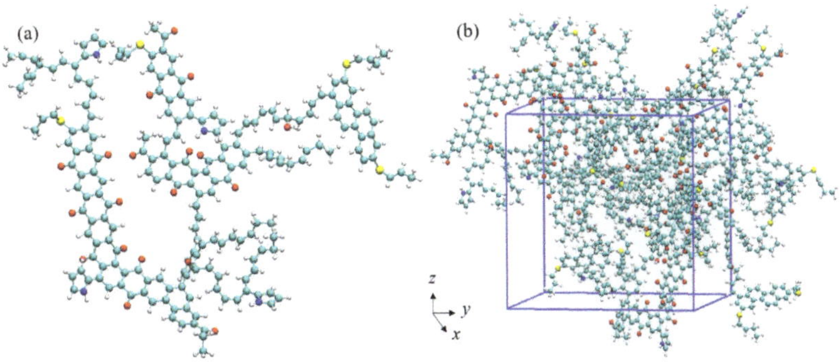

Figure 1. Snapshots of the unit structure of kerogen (**a**). Three-dimensional model (**b**). The atoms of C, H, O, N, and S are shown in gray, white, red, yellow, and blue, respectively.

Firstly, the initial three-dimensional model was optimized. Subsequently, an annealing simulation was employed to achieve an optimized model. Throughout the molecular mechanics and annealing molecular dynamics optimization, the total energy of the model underwent gradual changes. As a result, the final dimensions of the three-dimensional model were established at 3.184 nm × 3.184 nm × 3.184 nm. A visual representation of the three-dimensional model is depicted in Figure 1b. Throughout the molecular dynamics optimization process, the density of kerogen exhibited fluctuations, hovering around 1.035 g/cm^3. This value is close to the density of 1.01 g/cm^3 reported in a previous study once energy stabilization was achieved [37].

2.2. Simulation Details

COMPASS (condensed-phase optimized molecular potentials for atomistic simulation studies), widely employed in previous studies for organics, was chosen for the models [38]. To investigate the adsorption behavior of pure CH_4, CO_2, and N_2, along with their binary mixtures at different mole fractions in the nanopores of kerogen, GCMC was employed. The interactions between gas molecules and kerogen are governed by Van der Waals and electrostatic interactions [39]. The Metropolis sampling method was applied to determine the formation, migration, and rotation of molecules. Electrostatic long-range effects were calculated using the Ewald summation method, while Van der Waals interactions were handled with an atom-based method. First, the steepest descent algorithm was used to obtain the energy-minimized stable configuration. Then, the adsorption simulation was conducted under constant pressure and temperature conditions, with a total of 3×10^7 simulation steps. The first half of the simulation process ensured adsorption equilibrium, while the latter half focused on ensemble averaging and calculating parameters such as the adsorption capacity, adsorption energy, and heat of adsorption. The cutoff distance was set at 1.25 nm. The simulated temperatures were set to 303.15 K, 333.15 K, 348.15 K, and 363.15 K. The maximum simulated pressure was 30 MPa, and the adsorption and diffusion simulations were performed point by point to account for the actual temperature and pressure conditions of the shale reservoir. The detailed composition, temperature, and pressure conditions are shown in Table 1.

Table 1. Composition, temperature, and pressure conditions of kerogen system.

System	Temperature	Pressure
Pure CH_4	303.15 K 333.15 K 348.15 K 363.15 K	5 MPa, 10 MPa, 15 MPa, 20 MPa, 25 MPa, 30 MPa
Pure CO_2	333.15 K	
Pure N_2	333.15 K	
CH_4:CO_2 = 0.2:0.8 CH_4:CO_2 = 0.4:0.6 CH_4:CO_2 = 0.5:0.5 CH_4:CO_2 = 0.6:0.4 CH_4:CO_2 = 0.8:0.2	333.15 K	2.5 MPa, 5 MPa, 10 MPa, 15 MPa, 20 MPa, 25 MPa, 30 MPa
CH_4:N_2 = 0.2:0.8 CH_4:N_2 = 0.4:0.6 CH_4:N_2 = 0.5:0.5 CH_4:N_2 = 0.6:0.4 CH_4:N_2 = 0.8:0.2	333.15 K	2.5 MPa, 5 MPa, 10 MPa, 15 MPa, 20 MPa, 25 MPa, 30 MPa

A molecular dynamics simulation was employed to calculate the position, velocity, and energy of particles at various simulation times. Utilizing the configuration obtained from the adsorption simulations, the molecular dynamics method was applied to simulate the diffusion of shale gas within the kerogen system. The gas diffusion process used the same force field parameters and interaction forces between gases as those employed in the adsorption process. The adsorption configuration was employed for isothermal–isobaric simulations, with initial velocities randomized. The simulation utilized the Nosé–Hoover thermostat and Parrinello–Rahman barostat with temporal constants set to 1.0 ps and 4.0 ps, respectively. The total simulation time was set to 2 ns, with a time step of 1 fs, ensuring that the mean squared displacement exhibited a genuinely linear trend over time. Periodic boundary conditions were maintained throughout the entire simulation.

3. Results and Discussion

3.1. Pore Volume and Special Surface Area

The pore structure of kerogen significantly influences the adsorption and storage of shale gas. To quantify the pore volume and specific surface area, a three-dimensional model with probe molecules (CH_4, CO_2, and N_2) was analyzed [40,41]. The molecular diameters of CH_4, CO_2, and N_2 are 3.80 Å, 3.30 Å, and 3.64 Å, respectively. The resulting pore volume and specific surface area of the kerogen model with different probe molecules are presented in Table 2.

Table 2. Pore volume and special surface area of kerogen model with different probe molecules.

Probe Molecule	Pore Volume Å^3/uc	Special Surface Area Å^2/uc
CH_4	7086.27	4379.37
CO_2	8285.23	4881.91
N_2	7459.88	4549.75

3.2. Gas Fugacity

Gas fugacity represents the effective pressure exerted by a gas in a non-ideal system. It is used to account for the deviations of real gases from ideal gas behavior. The Peng–Robinson (PR) equation of state [42] was employed to calculate the fugacity at different temperatures and pressures, as illustrated in Figure 2. The critical parameters, including critical temperature, critical pressure, critical volume, acentric factor, and molar mass, are detailed in Table 3.

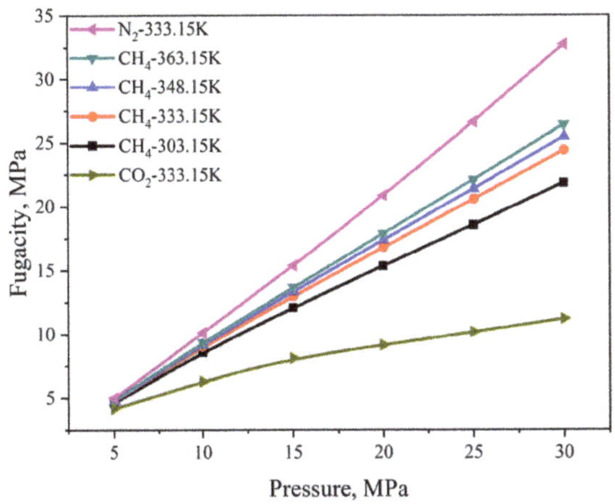

Figure 2. Fugacity of CH_4, CO_2, and N_2 under various temperature and pressure conditions.

Table 3. Critical parameters of CH_4, CO_2, and N_2 [43].

Critical Parameters	CH_4	CO_2	N_2
Critical temperature/K	190.560	304.140	126.190
Critical pressure/MPa	4.599	7.377	3.396
Critical volume/cm^3	98.600	94.340	90.100
Acentric factor KJ/(Kg°C)	0.011	0.224	0.037
Molar mass/g/mol	16.040	44.010	28.010

3.3. Adsorption Isotherm of Pure Component

The adsorption curves of CH_4, CO_2, and N_2 in the kerogen model were fitted using the Langmuir adsorption model [44], as depicted in Equation (1).

$$V = \frac{V_L P}{P_L + P} \tag{1}$$

where V is the adsorption capacity; V_L is the Langmuir volume, which reflects the maximum adsorption capacity; P_L is the Langmuir pressure, indicating the pressure for the half-maximal adsorption capacity; and P is the equilibrium pressure.

The adsorption isotherms of CH_4, CO_2, and N_2 in the kerogen model are illustrated in Figure 3. As observed in Figure 3, the adsorption capacity of different gases increased with the rise in pressure, with CO_2 exhibiting a significantly larger adsorption rate and capacity compared to CH_4 and N_2. The adsorption capacity approached saturation as the pressure reached 30 MPa. The descending order of adsorption capacity for gases was found to be $CO_2 > CH_4 > N_2$. Moreover, the adsorption capacity decreased with an increase in temperature under constant pressure conditions. This suggests that elevated temperatures negatively impact gas adsorption. The rationale behind this observation lies in the increased thermal motion of gas molecules with rising temperatures, leading to a higher average kinetic energy of gas molecules. This heightened kinetic energy may facilitate an easier escape from the kerogen surface, consequently reducing the gas adsorption capacity. In addition, the molecular diameter of CO_2 (approximately 3.3 Å) is slightly smaller than that of CH_4 (approximately 3.8 Å), which makes CO_2 molecules more easily able to enter and diffuse within the nanopores of kerogen. The smaller molecular diameter increases the opportunities for CO_2 to come into contact with the adsorbent surface, thereby enhancing adsorption capacity. Meanwhile, the CO_2 molecule is linear, whereas CH_4 is a tetrahedral molecule. The shape of the CO_2 molecule makes it easier to find suitable positions for adsorption in the narrow pores of the kerogen structure. Through a comparison between the simulation results and the Langmuir model fitting outcomes, the Langmuir adsorption parameters (V_L and P_L) and the average relative errors were determined and are presented in Table 4. Table 4 indicates that the Langmuir adsorption model achieved a fitting precision exceeding 0.99, with fitting errors less than 1.3%. These results affirm that the adsorption properties of gas in kerogen adhere to the Langmuir adsorption law.

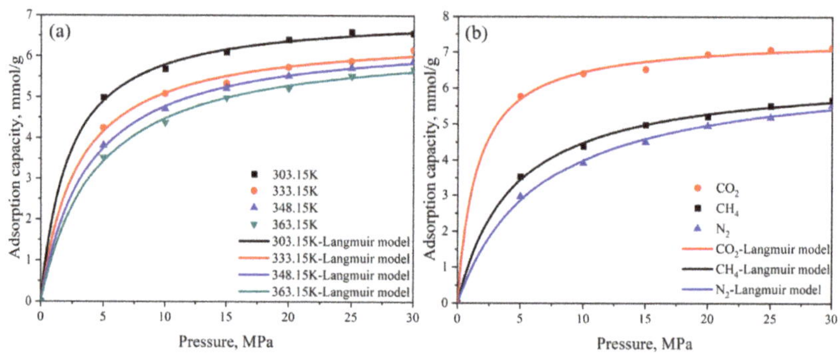

Figure 3. The adsorption capacity of CH_4 under different temperatures (**a**) and a comparison of CH_4, CO_2, and N_2 at 333.15 K (**b**).

Table 4. Fitting parameters (V_L and P_L) and errors of the Langmuir model.

Parameters	CH_4				CO_2	N_2
	303.15 K	333.15 K	348.15 K	363.15 K	333.15 K	333.15 K
V_L (mmol/g)	7.048	6.569	6.548	6.421	7.413	6.576
P_L (MPa)	2.170	2.885	3.704	4.329	1.504	6.480
R^2	0.9988	0.9972	0.9994	0.9986	0.9980	0.9979
ARE	1.02%	1.3%	0.66%	0.89%	0.85%	1.26%

Note: The R^2 value shows the goodness of fit, and *ARE* is the average relative error.

3.4. Isosteric Heat of Gas Adsorption

Figure 4a illustrates the variation in the isosteric heat of CH_4 adsorption under different pressure and temperature conditions. The observed trend indicates that the isosteric heat of CH_4 adsorption decreases with an increase in temperature. This suggests that the interaction between CH_4 molecules and kerogen diminishes as the temperature rises, leading to a decrease in the adsorption capacity of CH_4 molecules. In Figure 4b, a comparison of the isosteric heat of CH_4, CO_2, and N_2 adsorption under different pressures is presented. It is evident that the isosteric heat of CO_2 adsorption is significantly higher than that of CH_4 and N_2. Furthermore, the values of the isosteric heat of CH_4, CO_2, and N_2 adsorption are all less than 42 kJ/mol [45], indicating that the adsorption of gas molecules in kerogen systems falls under the category of physical adsorption.

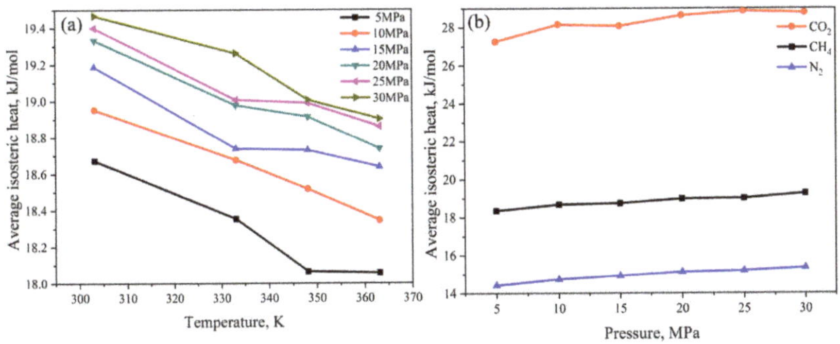

Figure 4. Isosteric heat of adsorption CH_4 under different temperature and pressure conditions (**a**) and comparison of CH_4, CO_2, and N_2 at 333.15 K (**b**).

3.5. Energy Distribution

Figure 5 illustrates the energy distribution curves of CO_2, CH_4, and N_2 in kerogen models. The curves display a single-peak distribution, with the maximum energy distribution values for CO_2, CH_4, and N_2 being −26.96 kJ/mol, −17.35 kJ/mol, and −13.17 kJ/mol, respectively. This result indicates that CO_2, CH_4, and N_2 primarily adsorb onto distinct sites within the kerogen models. In kerogen pores, CO_2, CH_4, and N_2 molecules compete for adsorption space and sites. Notably, the energy distribution range of CO_2 is considerably broader than that of CH_4 and N_2. This difference leads to a reduction in adsorption space and sites for CH_4 and N_2, consequently resulting in a decrease in the adsorption capacity of CH_4 and N_2. Additionally, it suggests that the adsorption of N_2 on kerogen is not as stable as that of CH_4 and CO_2.

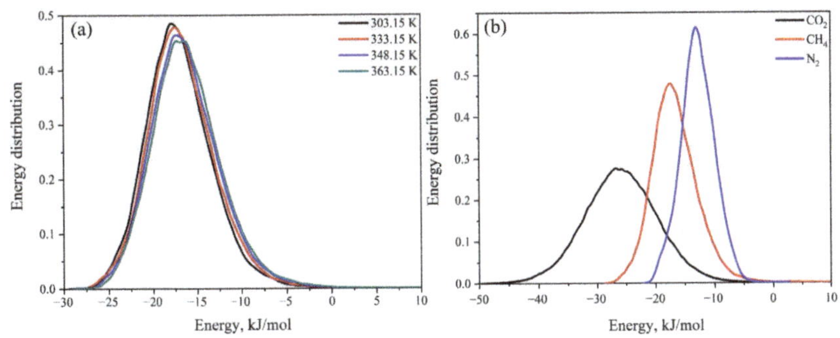

Figure 5. The energy distribution of CH_4 under different temperatures (**a**) and a comparison of CH_4, CO_2, and N_2 at 333.15 K (**b**).

3.6. Adsorption Isotherm of Binary Components

Figure 6 depicts the adsorption capacity of CH_4, CO_2, and N_2 within the nanopores of kerogen with varying mole fractions of CO_2 and N_2. At low pressures, the adsorption capacity of CH_4, CO_2, and N_2 increases with increasing pressure. However, at 10 MPa, the adsorption of CH_4 shows minimal changes with further pressure increases, indicating a negligible impact of pressure on CH_4 adsorption. Similarly, at 5 MPa and 20 MPa, the adsorption of CO_2 and N_2, respectively, exhibits minimal changes with further pressure increases. Under identical temperature and pressure conditions, the adsorption of CH_4 increases with the increasing mole fraction of CH_4, suggesting that the adsorption of CH_4 decreases with an increase in the content of N_2 (or CO_2). This observation indicates that the lower the mole fraction of CH_4 in the binary component mixture, the lower the CH_4 adsorption in kerogen pores. Furthermore, the adsorption capacity of CH_4 in a CH_4 and N_2

mixed system is significantly higher than that in a CH_4 and CO_2 mixed system, indicating that the adsorption capacity of CH_4 is greater than that of N_2 but less than that of CO_2.

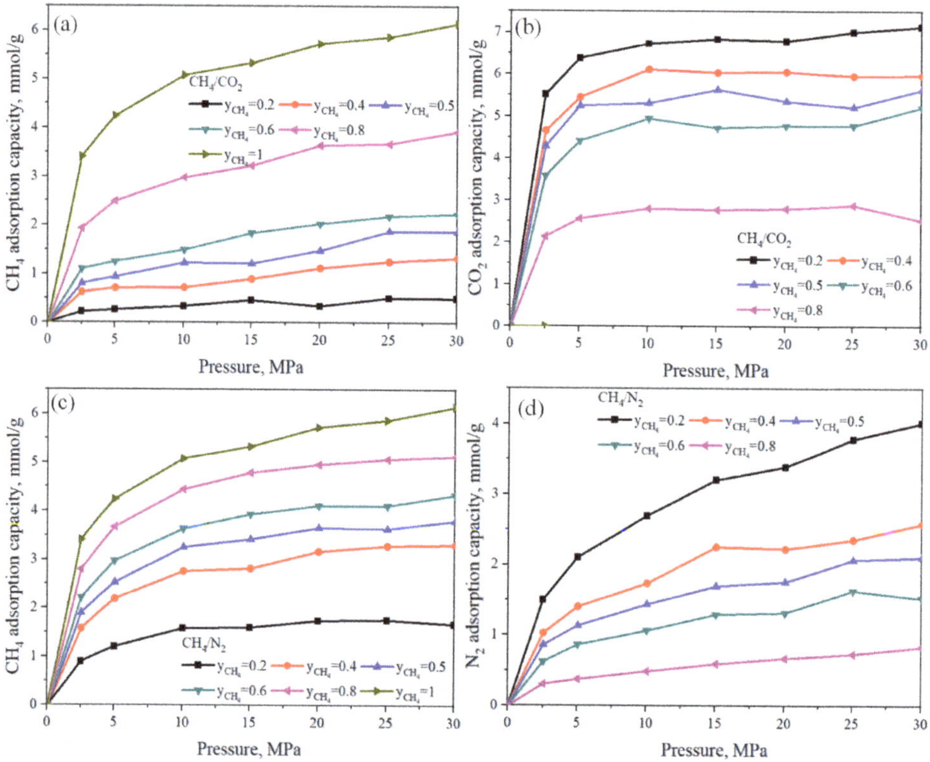

Figure 6. Adsorption capacities of CH_4 (**a**) and CO_2 (**b**) in nanopores of kerogen at different mole fractions of CH_4/CO_2. Adsorption capacities of CH_4 (**c**) and N_2 (**d**) in nanopores of kerogen at different mole fractions of CH_4/N_2.

3.7. Self-Diffusion Coefficient

The diffusion behaviors of shale gas in kerogen nanopores are characterized by the self-diffusion coefficient, calculated using the Einstein equation (Equation (2)) [46].

$$D_s = \lim_{t\to\infty}\frac{1}{6Nt}\left\langle \sum_{i=1}^{N} |r_i(t) - r_i(0)|^2 \right\rangle \quad (2)$$

where D_s is the self-diffusion coefficient, N is the number of particles, t is the time interval, $r(t)$ is the position of the gas molecule at time t, and $\left\langle |r_i(t) - r_i(0)|^2 \right\rangle$ is the mean square displacement (MSD) of molecules. The relationship between MSD and time is typically fitted to a straight line, and the self-diffusion coefficient is proportional to 1/6 of the slope of MSD versus time.

In Figure 7, it is evident that the self-diffusion coefficient of gas molecules decreases with an increase in pressure. At lower pressures, the reduction in the self-diffusion coefficient is relatively pronounced, but as pressure increases further, the decrease becomes more gradual. This observation is attributed to elevated pressure causing an increased number of gas molecules in the kerogen nanopores and a higher probability of collisions between molecules. The increasing motion resistance due to reduced available space results in a decrease in the self-diffusion coefficient. Beyond a certain pressure threshold, the velocity of gas molecules stabilizes gradually, leading to a less pronounced decline in the diffusion coefficient. With an increase in temperature, the irregular motion of gas molecules intensifies, resulting in a larger self-diffusion coefficient. Consequently, the interaction between gas molecules becomes the primary factor controlling molecular diffusion. At the same temperature, the self-diffusion coefficient of fluids follows the descending order: N_2 > CH_4 > CO_2. This trend can be explained by classical diffusion theory, where the diffusion coefficient is inversely proportional to the square root of the molecular mass. CO_2 molecules have a large mass and hence a smaller self-diffusion coefficient. However, at the same temperature, CH_4, CO_2, and N_2 possess the same average kinetic energy. Due to their lighter mass, N_2 molecules move faster in the nanopores of kerogen, resulting in a higher diffusion rate for N_2. Moreover, smaller N_2 molecules can more easily pass through narrow pores, whereas larger CH_4 and CO_2 molecules may encounter restrictions. Additionally, lighter N_2 molecules regain their velocity faster after collisions, thus diffusing more rapidly within the nanopores of kerogen. This behavior is attributed to stronger interactions between CO_2 molecules and kerogen compared to CH_4, which reduces the movement of CO_2 molecules and thereby lowers their diffusion coefficient.

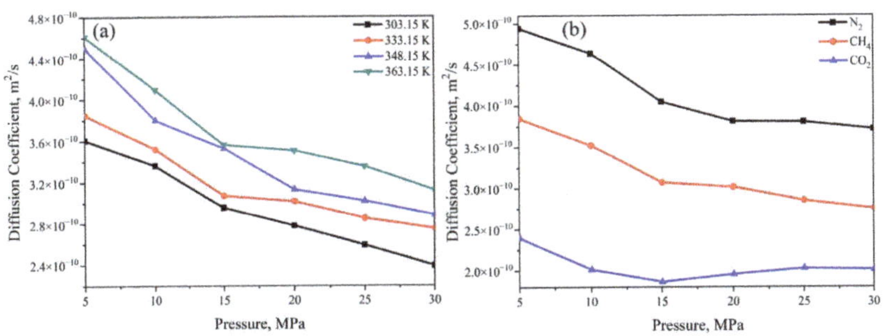

Figure 7. The self-diffusion coefficient of CH_4 under different temperatures (**a**) and a comparison of CH_4, CO_2, and N_2 at 333.15 K (**b**).

3.8. Radial Distribution Function

The radial distribution function is employed to characterize the distribution of gas molecules around kerogen structures and assess their interactions with kerogen. The radial distribution function is defined as the ratio of local density to average bulk density [47], and it is commonly employed to characterize the microstructure of particles. In Figure 8, the radial distribution functions of gas molecules are compared with atoms (C, H, O, N, and S) in kerogen. The peak values between gas molecules and O, N, and C atoms in kerogen are notably larger than those between gas molecules and H and S atoms, indicating weaker interactions between gas molecules and H and S atoms. A significant peak is observed between gas molecules and N atoms, suggesting the presence of N-H bonds in kerogen, which leads to stronger interactions between gas molecules and N-H bonds. Furthermore, the first peak of radial distribution functions between CO_2 and O occurs at 0.335 nm, and it is sharper and larger than those between CH_4-O and CH_4-N, indicating strong adsorption of CO_2 on sites with oxygen-containing functional groups. This observation is consistent with the snapshots of gas molecule distributions on the surface of local kerogen structures (Figure 8d–f), suggesting the superior adsorption capability of CO_2 in kerogen, significantly exceeding that of CH_4 and N_2.

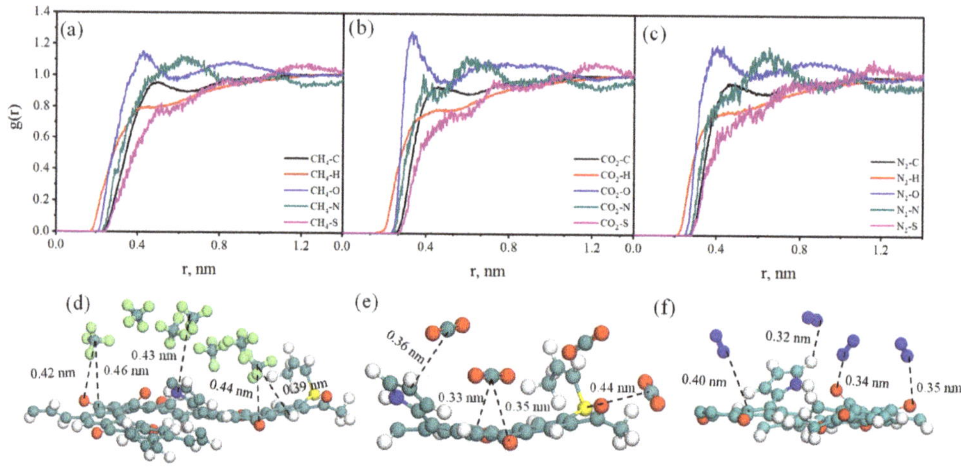

Figure 8. Radial distribution function between CH_4 (**a**), CO_2 (**b**), and N_2 (**c**) and various atoms of kerogen. Snapshots showing distribution of CH_4 (**d**), CO_2 (**e**), and N_2 (**f**) molecules on surface of local kerogen structure.

3.9. Distribution of CH_4, CO_2, and N_2 in Kerogen Models

The distribution characteristics of CH_4, CO_2, and N_2 in the kerogen model were analyzed. Figure 9 shows the snapshots of pure CH_4 (a), CO_2 (b), and N_2 (c) distributions in the nanopores of kerogen, demonstrating that gas molecules predominantly occupy the nanopores of the kerogen model. This observation provides a visual explanation for the previously noted trend in adsorption capacity, where, under the same temperature and pressure conditions, the adsorption capacity of gases in the kerogen nanopores follows the order $CO_2 > CH_4 > N_2$. Figure 9d–i depict the distribution of binary components (CH_4/CO_2 and CH_4/N_2) at three different mole fractions in the nanopores of kerogen. The results reveal that as the mole fraction of CH_4 increases, the number of CH_4 molecules in kerogen nanopores also increases. However, even at high mole fractions of CH_4, CO_2 continues to occupy a significant portion of the available space. Specifically, when the mole fraction of the two components is 1:1 in the binary mixture, the adsorption amount of CO_2 is notably greater than that of CH_4. These findings indicate a stronger adsorption capacity of CO_2 compared to CH_4 in kerogen nanopores, with CO_2 exerting a more significant influence on the adsorption behavior of CH_4. Similarly, the results demonstrate that the adsorption capacity of N_2 in kerogen nanopores is lower than that of CH_4. CO_2 molecules that are injected into a shale gas reservoir can competitively adsorb with CH_4 in kerogen nanopores, thereby promoting the release of CH_4 molecules. This process effectively enhances the recovery rate of shale gas. While CH_4 adsorption is relatively stable, theoretical approaches can be explored to partially desorb CH_4 through stronger additives or optimized displacement conditions to minimize its retention in kerogen. In summary, understanding the differential adsorption mechanisms of CO_2 and CH_4 in kerogen nanopores is crucial for optimizing extraction strategies, thereby improving the efficiency and economics of shale gas recovery.

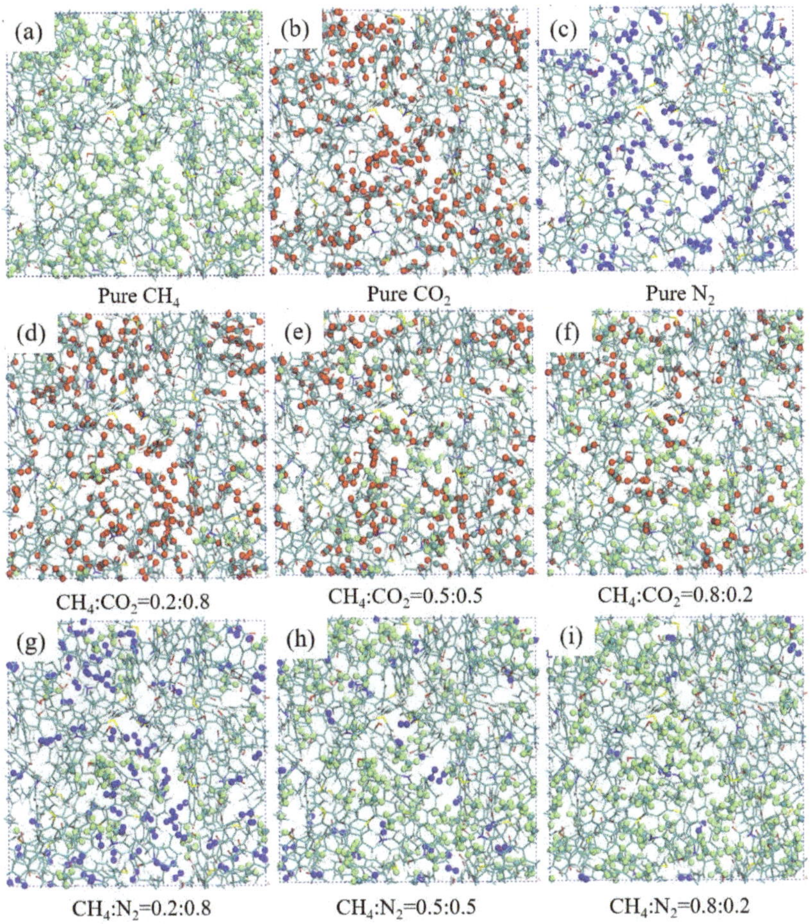

Figure 9. Snapshots of pure CH_4 (**a**), CO_2 (**b**), and N_2 (**c**) and binary component (CH_4/CO_2 and CH_4/N_2) distributions in nanopores of kerogen, illustrating three different CH_4 mole fractions (**d**–**i**).

4. Conclusions

The adsorption and diffusion of CH_4, CO_2, and N_2, including their binary mixtures with varying mole fractions in kerogen nanopores, were investigated using GCMC and MD methods. The results exhibit a strong correlation (0.99) with the Langmuir adsorption model. Gas adsorption capacity shows an increase with pressure and a decrease with temperature. Specifically, the adsorbed amount of CH_4 decreases as the mole fraction of CO_2 and N_2 in the gas phase increases. Below 15 MPa, the adsorption capacity of all gas molecules increases rapidly. N_2 exhibits adsorption at higher energy sites on kerogen, resulting in reduced adsorption stability and lower capacity compared to CH_4 and CO_2. The presence of oxygen-containing functional groups enhances gas adsorption. At identical temperature and pressure conditions, the diffusion capacity of gas follows the order $N_2 > CH_4 > CO_2$, ranging from 2×10^{-10} m^2/s to 5×10^{-10} m^2/s. These findings contribute to understanding the mechanisms of gas storage in kerogen nanopores, particularly in advancing strategies for the CO_2 displacement of CH_4 in shale gas extraction.

Author Contributions: Conceptualization, Y.L.; Software, X.L.; Investigation, X.L., Q.Z., N.W., Y.Q., Y.Z. and C.Y.; Resources, S.X., Q.Z., N.W., Y.Q. and Y.Z.; Data curation, C.Y.; Writing—original draft, S.X.; Writing—review & editing, Y.L.; Supervision, Y.L. All authors have read and agreed to the published version of the manuscript.

Funding: This work is supported by the Shenzhen Science and Technology Program (RCBS20221008093233048), Guangdong Basic and Applied Basic Research Foundation (2022A1515010093), Center for Computational Science and Engineering at Southern University of Science and Technology.

Data Availability Statement: The data presented in this study are available on request from the corresponding author.

Conflicts of Interest: Author Shouheng Xiao, Qiang Zheng, Ning Wang, Youyin Zhang and Chuanjun Yi were employed by the company Research Institute of Exploration & Development, Xinjing Oilfield Company. Author Yun Qiao was employed by the company Research Institute of Engineering Technology, Xinjing Oilfield Company. The remaining authors declare that the research was conducted in the absence of any commercial or financial relationships that could be construed as a potential conflict of interest.

Nomenclature

GCMC	Grand canonical Monte Carlo
MD	Molecular dynamics
RDF	Radial distribution function
MSD	Mean square displacement
D_s	Self-diffusion coefficient (m^2/s)
N	Number of particles
t	Time interval (s)
$r_i(t)$	Position of the *i*-th gas molecule at time t
$\left\langle \lvert r_i(t) - r_i(0) \rvert^2 \right\rangle$	Mean square displacement of molecules
V	Adsorption capacity
V_L	Langmuir volume, the maximum adsorption capacity
P_L	Langmuir pressure, the pressure for half-maximal adsorption capacity
P	Equilibrium pressure
T	Temperature (K)
ρ(r)	Local density
ρ0	Average bulk density
CH_4	Methane
CO_2	Carbon dioxide
N_2	Nitrogen
T_c	Critical temperature (K)
P_c	Critical pressure (MPa)
V_c	Critical volume (m^3/mol)
w	Acentric factor
M	Molar mass (g/mol)
ΔH	Isosteric heat of adsorption (kJ/mol)

References

1. Wang, Q.; Chen, X.; Jha, A.N.; Rogers, H. Natural gas from shale formation—The evolution, evidences and challenges of shale gas revolution in United States. *Renew. Sustain. Energy Rev.* **2014**, *30*, 1–28. [CrossRef]
2. Etminan, S.R.; Javadpour, F.; Maini, B.B.; Chen, Z. Measurement of gas storage processes in shale and of the molecular diffusion coefficient in kerogen. *Int. J. Coal Geol.* **2014**, *123*, 10–19. [CrossRef]
3. Mcglade, C.; Speirs, J.; Sorrell, S. Unconventional gas—A review of regional and global resource estimates. *Energy* **2013**, *55*, 571–584. [CrossRef]
4. Hughes, J.D. Energy: A reality check on the shale revolution. *Nature* **2013**, *494*, 307–308. [CrossRef]
5. Gasparik, M.; Bertier, P.; Gensterblum, Y.; Ghanizadeh, A.; Krooss, B.M.; Littke, R. Geological controls on the methane storage capacity in organic-rich shales. *Int. J. Coal Geol.* **2014**, *123*, 34–51. [CrossRef]
6. Fang, H.; Zou, H.; Lu, Y. Mechanisms of shale gas storage: Implications for shale gas exploration in China. *AAPG Bull.* **2013**, *97*, 1325–1346.

7. Zhang, T.; Ellis, G.S.; Ruppel, S.C.; Milliken, K.; Yang, R. Effect of organic-matter type and thermal maturity on methane adsorption in shale-gas systems. *Org. Geochem.* **2012**, *47*, 120–131. [CrossRef]
8. Curtis, J.B. Fractured shale-gas systems. *AAPG Bull.* **2002**, *86*, 1921–1938.
9. Montgomery, S.L.; Jarvie, D.M.; Bowker, K.A.; Pollastro, R.M. Mississippian Barnett Shale, Fort Worth basin, north-central Texas: Gas-shale play with multi–trillion cubic foot potential. *AAPG Bull.* **2005**, *90*, 963–966. [CrossRef]
10. Liu, D.; Yuan, P.; Liu, H.; Li, T.; Tan, D.; Yuan, W.; He, H. High-pressure adsorption of methane on montmorillonite, kaolinite and illite. *Appl. Clay Sci.* **2013**, *85*, 25–30. [CrossRef]
11. Duan, S.; Gu, M.; Du, X.; Xian, X. Adsorption Equilibrium of CO_2 and CH_4 and Their Mixture on Sichuan Basin Shale. *Energy Fuels* **2016**, *30*, 2248–2256. [CrossRef]
12. Hu, H.; Zhang, T.; Wiggins-Camacho, J.D.; Ellis, G.S.; Lewan, M.D.; Zhang, X. Experimental investigation of changes in methane adsorption of bitumen-free Woodford Shale with thermal maturation induced by hydrous pyrolysis. *Mar. Pet. Geol.* **2015**, *59*, 114–128. [CrossRef]
13. Wang, Z.; Li, Y.; Guo, P.; Meng, W. Analyzing the Adaption of Different Adsorption Models for Describing the Shale Gas Adsorption Law. *Chem. Eng. Technol.* **2016**, *39*, 1921–1932. [CrossRef]
14. Ji, L.; Zhang, T.; Milliken, K.L.; Qu, J.; Zhang, X. Experimental investigation of main controls to methane adsorption in clay-rich rocks. *Appl. Geochem.* **2012**, *27*, 2533–2545. [CrossRef]
15. Guo, S. Experimental study on isothermal adsorption of methane gas on three shale samples from Upper Paleozoic strata of the Ordos Basin. *J. Pet. Sci. Eng.* **2013**, *110*, 132–138. [CrossRef]
16. Luo, X.; Wang, S.; Wang, Z.; Jing, Z.; Lv, M.; Zhai, Z.; Han, T. Adsorption of methane, carbon dioxide and their binary mixtures on Jurassic shale from the Qaidam Basin in China. *Int. J. Coal Geol.* **2015**, *150*, 210–223. [CrossRef]
17. Bi, H.; Jiang, Z.; Li, J.; Li, P.; Chen, L.; Pan, Q.; Wu, Y. The Ono–Kondo model and an experimental study on supercritical adsorption of shale gas: A case study of on Longmaxi shale in southeastern Chongqing, China. *J. Nat. Gas Sci. Eng.* **2016**, *35*, 114–121. [CrossRef]
18. Chareonsuppanimit, P.; Mohammad, S.A.; Robinson, R.L., Jr.; Gasem, K.A. High-pressure adsorption of gases on shales: Measurements and modeling. *Int. J. Coal Geol.* **2012**, *95*, 34–46. [CrossRef]
19. Kim, C.; Jang, H.; Lee, J. Experimental investigation on the characteristics of gas diffusion in shale gas reservoir using porosity and permeability of nanopore scale. *J. Pet. Sci. Eng.* **2015**, *133*, 226–237. [CrossRef]
20. Liu, G.; Zhao, Z.; Sun, M.; Li, J.; Hu, G.; Wang, X. New insights into natural gas diffusion coefficient in rocks. *Pet. Explor. Dev.* **2012**, *39*, 597–604. [CrossRef]
21. Javadpour, F. Nanopores and Apparent Permeability of Gas Flow in Mudrocks (Shales and Siltstone). *J. Can. Pet. Technol.* **2009**, *48*, 16–21. [CrossRef]
22. Wu, K.; Li, X.; Wang, C.; Yu, W.; Chen, Z. Model for Surface Diffusion of Adsorbed Gas in Nanopores of Shale Gas Reservoirs. *Ind. Eng. Chem. Res.* **2015**, *54*, 3225–3236. [CrossRef]
23. Rezlerová, E.; Brennan, J.K.; Lísal, M. Methane and carbon dioxide in dual-porosity organic matter: Molecular simulations of adsorption and diffusion. *AICHE J.* **2021**, *67*, e16655. [CrossRef]
24. Zhang, C.; Yao, Y.; Swennen, R.; Zhang, Y. Re-evaluating the methane adsorption behavior in shale kerogen: Unifying experiment and molecular simulation. *Phys. Fluids* **2024**, *36*, 022002. [CrossRef]
25. Tesson, S.; Firoozabadi, A. Methane adsorption and self-diffusion in shale kerogen and slit nanopores by molecular simulations. *J. Phys. Chem. C* **2018**, *122*, 23528–23542. [CrossRef]
26. Dawass, N.; Vasileiadis, M.; Peristeras, L.D.; Papavasileiou, K.D.; Economou, I.G. Prediction of Adsorption and Diffusion of Shale Gas in Composite Pores Consisting of Kaolinite and Kerogen using Molecular Simulation. *J. Phys. Chem. C* **2023**, *127*, 9452–9462. [CrossRef]
27. Yuan, S.; Gang, H.-Z.; Liu, Y.-F.; Zhou, L.; Irfan, M.; Yang, S.-Z.; Mu, B.-Z. Adsorption and Diffusion Behaviors of CO_2 and CH_4 Mixtures in Different Types of Kerogens and Their Roles in Enhanced Energy Recovery. *Sustainability* **2022**, *14*, 14949. [CrossRef]
28. Wang, S.; Feng, Q.; Javadpour, F.; Yang, Y.-B. Breakdown of fast mass transport of methane through calcite nanopores. *J. Phys. Chem. C* **2016**, *120*, 14260–14269. [CrossRef]
29. Li, Y.; Chen, M.; Tang, H.; Han, S.; Song, H.; Wang, P.; Zhao, Y.; Zhu, J. Insights into Carbon Dioxide Hydrate Nucleation on the External Basal Surface of Clay Minerals from Molecular Dynamics Simulations. *ACS Sustain. Chem. Eng.* **2022**, *10*, 6358–6369. [CrossRef]
30. Li, Y.; Chen, M.; Liu, C.; Song, H.; Yuan, P.; Zhang, B.; Liu, D.; Du, P. Effects of Layer-charge Distribution of 2:1 Clay Minerals on Methane Hydrate Formation: A Molecular Dynamics Simulation Study. *Langmuir* **2020**, *36*, 3323–3335. [CrossRef] [PubMed]
31. Li, Y.; Chen, M.; Song, H.; Yuan, P.; Liu, D.; Zhang, B.; Bu, H. Methane Hydrate Formation in the Stacking of Kaolinite Particles with Different Surface Contacts as Nanoreactors: A Molecular Dynamics Simulation Study. *Appl. Clay Sci.* **2020**, *186*, 105439. [CrossRef]
32. Collell, J.; Ungerer, P.; Galliero, G.; Yiannourakou, M.; Montel, F.; Pujol, M. Molecular Simulation of Bulk Organic Matter in Type II Shales in the Middle of the Oil Formation Window. *Energ. Fuel* **2014**, *28*, 7457–7466. [CrossRef]
33. Ungerer, P.; Collell, J.; Yiannourakou, M. Molecular Modeling of the Volumetric and Thermodynamic Properties of Kerogen: In fluence of Organic Type and Maturity. *Energ. Fuel* **2015**, *29*, 91–105. [CrossRef]
34. Yao, S.; Jiao, K.; Li, M. Advances in research of coal and kerogen nanostructure. *Adv. Earth Sci.* **2012**, *27*, 367–378.

35. Liu, X.; Luo, D.; Xiong, J.; Liang, L. Construction of the average molecular modeling of the kerogen from Longmaxi formation *Chem. Ind. Eng. Prog.* **2017**, *36*, 530–537.
36. Wang, Z.H.; Li, Y.; Meng, W.J.; Guo, P.; Luo, Q.; Ran, Z.L. Density functional theory study on the adsorption of methane on kerogen with different functional groups. *Appl. Ecol. Env. Res.* **2017**, *15*, 861–869. [CrossRef]
37. Wang, Q.; Pan, S.; Bai, J.; Chi, M.; Cui, D.; Wang, Z.; Liu, Q.; Xu, F. Theoretical study of structural and spatial properties of kerogen. *Energ. Fuel* **2019**, *33*, 9559–9569. [CrossRef]
38. Sun, H.; Ren, P.; Fried, J.R. The COMPASS Forcefield: Parameterization and Validation for Polyphosphazenes. *Comput. Theor Polym. Sci.* **1998**, *8*, 229. [CrossRef]
39. Mosher, K.; He, J.; Liu, Y.; Rupp, E.; Wilcox, J. Molecular simulation of methane adsorption in micro- and mesoporous carbons with applications to coal and gas shale systems. *Int. J. Coal Geol.* **2013**, *109*, 36–44. [CrossRef]
40. Connolly, M.L. Analytical molecular surface calculation. *J. Appl. Crystallogr.* **1983**, *6*, 548–558. [CrossRef]
41. Zeitler, T.R.; Allendorf, M.D.; Greathouse, J.A. Grand Canonical Monte Carlo Simulation of Low-Pressure Methane Adsorption in Nanoporous Framework Materials for Sensing Applications. *J. Phys. Chem. C* **2011**, *116*, 3492–3502. [CrossRef]
42. Peng, D.Y.; Robinson, D.B. A New Two-Constant Equation of State. *Ind. Eng. Chem. Fundam.* **1976**, *15*, 92–94. [CrossRef]
43. Smith, J.M.; Ness, H.C.V.H.C. *Introduction to Chemical Engineering Thermodynamics*; McGraw-Hill: New York, NY, USA, 2001.
44. Langmuir, I. The Adsorption of Gases on Plane Surfaces of Glass, Micaand and Platinum. *J. Chem. Phys.* **2015**, *40*, 1361–1403.
45. Fu, X.; Shen, W.; Yao, T. *Physical Chemistry*, 4th ed.; Higher Education: Beijing, China, 1993.
46. Allen, M.P.; Tildesley, D.J. *Computer Simulation of Liquids*; Clarendon: Oxford, UK, 1987.
47. Park, S.H.; Sposito, G. Monte Carlo Simulation of Total Radial Distribution Functions for Interlayer Water in Li-, Na-, and K-Montmorillonite Hydrates. *J. Steroid Biochem.* **2000**, *104*, 177–182. [CrossRef]

processes

MDPI

Article

Enhanced Study of CO_2 Hydrate Formation in Marine Oil–Gas Based on Additive Effect

Yang Ge [1,2], Haihong Chen [1,2], Rui Qin [1,2], Haiyuan Yao [1,2], Ting Huang [1,2], Xin Lv [3], Huiyong Liang [3] and Shi Shen [3,*]

1 State Key Laboratory of Offshore Natural Gas Hydrates, Beijing 100028, China; qinrui2@cnooc.com.cn (R.Q.); huangting7@cnooc.com.cn (T.H.)
2 China National Offshore Oil Corporation Research Institute Co., Ltd., Beijing 100028, China
3 Ningbo Institute of Dalian University of Technology, Ningbo 315016, China; lvxin@dlut.edu.cn (X.L.)
* Correspondence: shenshi_nbi@dlut.edu.cn

Abstract: During marine oil–gas extraction, significant amounts of carbon dioxide (CO_2) gas are often produced. Effectively separating these associated CO_2 gases during extraction has become a critical technical challenge. Therefore, this paper aims to enhance the efficiency of CO_2 hydrate-based capture technology and conduct relevant research. The goal is to increase the driving force for hydrate formation by combining the traditional thermodynamic additive TBAB with pressure modulation and to improve the hydrate formation rate through the use of multiple kinetic promoters. This paper presents the initial investigation into the effect of the thermodynamic accelerator tetrabutylammonium bromide (TBAB) on the characteristics of CO_2 hydrate formation. The promotion effects of TBAB solutions with varying mass concentrations (3%, 5%, 7.5%, 10%) and reaction pressures (3 MPa, 4 MPa) were subjected to a systematic analysis, and the optimal conditions were identified as 4 MPa and a 5 wt% TBAB concentration. Subsequently, the impact of combining TBAB with kinetic promoters (SDS, nano Al_2O_3, L-methionine, L-leucine) on CO_2 hydrate generation characteristics was further investigated. In this paper, the effect of a single promoter on the generation characteristics of CO_2 hydrate was investigated, and the efficient carbon trapping ability of the complex promoter was verified, which provides theoretical support for the application of CO_2 trapping technology using the hydrate method.

Keywords: flue gas; CO_2 hydrate; promoter; generation characteristics

1. Introduction

At present, global climate change has become a major challenge for the international community. In 2023, global CO_2 emissions reached 35.8 billion tons [1]. Among them, about 78% of CO_2 from flue gas emitted by fossil fuel combustion will become the maximum emissions source in 2022 [2]. The operation of internal combustion engines is also dependent on the combustion of fossil fuels, resulting in the emission of considerable quantities of CO_2 into the atmosphere [3,4]. Therefore, carbon capture technology has received increasing attention [5,6].

Conventional CO_2 capture technologies cover a variety of means [7–12], such as chemical absorption, physical adsorption, and low-temperature condensation separation. However, there are problems such as a large amount of absorbent used, high energy consumption, easy leakage and the corrosion of equipment, high investment cost of the device, residual impurities left over, as well as degradation and foaming of the absorbing medium, etc., so it is necessary to continue to push forward the research and development of new technologies that are more efficient, economical and environmentally friendly. Among the emerging technologies, the hydrate method shows great potential due to its low energy consumption, simple process, and good environmental compatibility, and is considered to be the frontier direction in the field of CO_2 capture [13–15]. Hydrate carbon capture and storage is a technology that converts CO_2 into hydrates and subsequently sequesters it. This method exploits the fact that CO_2 forms hydrates with water under high pressure and low temperature, thereby achieving gas separation and long-term storage of CO_2 [16]. Hydrates are solid cage-like structures formed by gas molecules and water molecules under specific conditions, which offer high CO_2 storage efficiency and low risk of leakage [17].

Gas hydrates, also known as cage hydrates, are solid crystals that form under specific high-pressure, low-temperature conditions [18–23]. They construct diverse cage-like structures through hydrogen bonding between water molecules and embed gas molecules in them to form ice-like substances [24–27]. As an emerging means of CO_2 capture, hydrate-based technology still faces challenges in terms of application and promotion [15,28]. For example, its formation and stabilization require specific high-pressure and low-temperature conditions, slow generation rates, and low gas storage density [25,29,30]. Currently, hydrate formation can be enhanced by adding hydrate promoters, which can be categorized into two main groups: thermodynamic promoters and kinetic promoters, according to their mechanism of action [16–19].

Thermodynamic promoters simplify the process of hydrate formation and regulate phase equilibrium by reducing the required energy threshold required for the reaction to occur. Among the common thermodynamic promoters are tetrahydrofuran (THF), cyclopentane (CP), tetrabutylammonium bromide (TBAB), tetrabutylammonium fluoride (TBAF), and tetrahydropyran (THP). TBAB and THF are effective and commonly used thermodynamic accelerators, which are able to generate hydrates by themselves and occupy only large cage-like structures, providing small cage sites for CO_2, thus enhancing CO_2 capture [31,32]. During the hydrate generation process, the presence of TBAB reduces the equilibrium pressure threshold of hydrate formation and enhances the separation efficiency of CO_2. Li et al. [33] verified the thermodynamic properties of TBAB by adding 5 wt% TBAB in a mixture of 19.9% CO_2 + 80% N_2 at a temperature of 277.5 K and a pressure of 4.3–7.3 MPa. Jean et al. [34] demonstrated that the addition of 0.29 mol% TBAB drastically reduced the pressure during CO_2/N_2 hydrate formation. Furthermore, TBAB is an environmentally friendly catalyst that is non-volatile, non-flammable, non-corrosive, and exhibits low cost and high thermochemical stability. Together, these findings demonstrate the practical application value of TBAB in the field of flue gas carbon capture. Tang [35] used 0.5 to 3.0 mol% aqueous THF solutions to capture CO_2 from a gas mixture of 59% CO_2 + 41% N_2, and CO_2 recoveries (0.71–0.85) were obtained that were significantly greater than those of the pure water system (0.54). Tetrabutylammonium fluoride (TBAF) had the same effect of lowering the equilibrium pressure and elevating the reaction temperature. Kim [36] et al. tested the generation characteristics of CO_2 hydrate by adding TBAF to a gas mixture (20% CO_2 + 80% N_2). TBAF significantly changed the phase equilibrium of the hydrate, and the phase equilibrium pressure was reduced to 1.04 MPa at a temperature of 300 K. However, the CO_2 capture ability was poor.

Kinetic promoters accelerate the rate of hydrate generation mainly by changing the kinetic properties of the solution, such as increasing the diffusion coefficient of the solution and decreasing the gas–liquid surface tension [37]. Such promoters cover several classes of compounds, mainly including nanoparticles [38,39] (alumina Al_2O_3, graphite powder, titanium dioxide), porous media [29] (e.g., quartz sand, silica gel), ionic liquids, as well as [40] bio-promoters (e.g., amino acids [41], lignosulfonates), surfactants [42], etc., and the common surfactants are SDS (sodium dodecyl sulfate), CTAB (brominated cetyltrimethylammonium), and STS (sodium tetradecyl sulfate) and so on. Said et al. [43] found that nano-Al_2O_3 showed significant enhancement effects at concentrations of 0.1 to 0.3 wt%, with gas consumption increasing by 30% to 65%. Liu et al. [40] investigated the significant effect of 0.1 wt% L-methionine on the absorption rate of CO_2 gas and the promotion of hydrate formation in 99.8% CO_2 gas. Li et al. [44] systematically investigated the role of amino acids, with a special focus on the effect of L-tryptophan as a kinetic enhancer of carbon dioxide hydrate generation. Yan Li et al. [45] found that 2 wt% L-leucine had a significant effect on the absorption rate of CO_2 gas as well as on the promotion of hydrate formation rate, with an optimal CO_2 normalized absorption rate of 62.17 ± 0.78 mmol/mol and a CO_2 occupancy of 66% in the hydrate phase.

As the pro-generation effect of hydrate promoters was verified, scholars began to investigate the effect of thermo–kinetic hybrid compounding additives on CO_2 hydrate generation, and Kumar [46] found that the process of CO_2 hydrate generation in a gas mixture of CO_2 and N_2 was positively affected in a solvent mixture of THF and SDS. Tang [35] also used SDS and THF as additives to capture CO_2 from a gas mixture of 59% CO_2 + 41% N_2 by the hydrate method and obtained that the most effective concentration range for SDS was 100 to 300 ppm.

Hydraulic fracturing is a technique employed in the extraction of gas hydrates, whereby fractures are formed at the base of the deposit, and a coolant with a temperature exceeding that of the deposit is pumped into these fractures. The principal objective of the method is the release of natural gas from the hydrate and the subsequent transportation of the decomposition products and the coolant to the surface. By applying high pressure inside the wellbore, the fracturing fluid can create cracks in the rock, increasing the flow path for the fluid and facilitating the release of natural gas.

The results of the above scholars show that single additives have a certain promotion effect on CO_2 hydrate generation, but there are fewer studies on the effect of thermo–kinetic hybrid additives on CO_2 hydrate generation, especially on changing the CO_2 hydrate generation characteristics to improve the capture effect in the associated gas and industrial waste gas. Therefore, in this paper, single additives that have been proven to have significant promotion effects were first screened for experiments, and evaluation criteria were set based on the experimental data. Then, the effects of additive combinations on CO_2 capture by the hydrate method are analyzed, and the optimal additive ratios under specific experimental conditions are determined. The most effective promoter combinations applicable to associated gas and industrial waste gas were subjected to analysis and determination through synthesis. As the quantity of quasi-liquid water represents the limiting factor for the formation of hydrates and the diffusion of gases within the system, the rate of hydrate formation may be reduced at lower temperatures. Furthermore, the rate of gas diffusion is also reduced at lower temperatures, which in turn increases the time required for gas molecules to come into contact with quasi-liquid water. Consequently, we selected higher temperatures and corresponding pressure conditions as the experimental conditions [47]. The characteristics and kinetic parameters of CO_2 hydrate generation were optimized.

2. Experimental Part

2.1. Experimental Set Up

The experimental system used to study the effect of accelerators on CO_2 capture by the hydrate method was built in this paper, and its structure is shown in Figure 1. The system mainly consists of a visible reactor (inner diameter of 70 mm, inner cavity height of 65 mm, volume of 250 mL), a magnetic stirrer, a constant temperature water bath, a Matsushima thermocouple (accuracy of 0.01 K), a Baoji Hengtong BP93420 pressure sensor (range 0~15 MPa, accuracy of 0.075%), and a data acquisition system from Advantech. In addition to the experimental setup, there were measuring instruments such as an electronic balance (OHAUS International Trading Co., Ltd., Shanghai, China, model CP513, maximum weighing capacity 510 g, dividing value 0.1 g) and a gas chromatograph (Tianmei Technology Co., Ltd., Beijing, China, model GC7900). In this experiment, 3 mm stainless steel tubing supplied by the Kumagawa manufacturer was used as the piping connecting the reactor, gas source, and detection sensor, and the valves were mainly ball valves and needle valves.

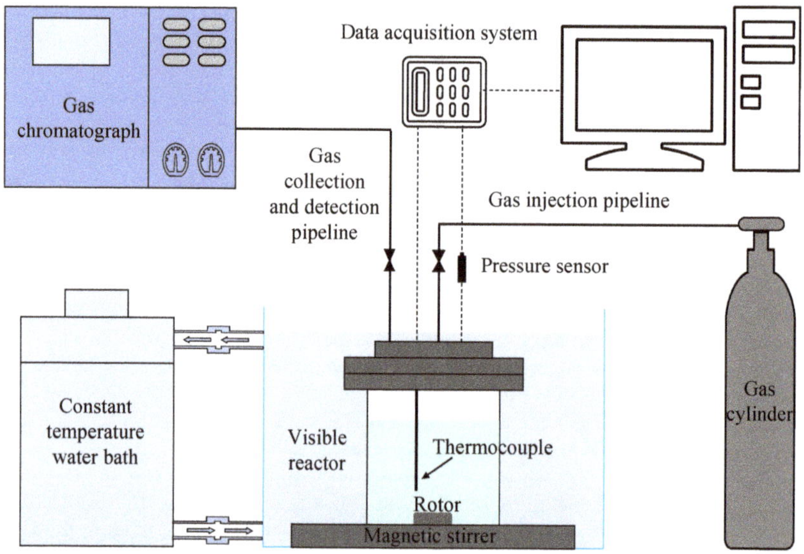

Figure 1. Schematic diagram of the experimental system.

2.2. Experimental Materials and Methods

2.2.1. Experimental Materials

The specifications of the accelerator and solvent materials used in this paper are shown in Table 1 below:

Table 1. Experimental material list.

Materials	Abbreviation/ Molecular Formula	Parameter	Source
Deionized water	H_2O		Lab-created
simulation Flue gas	$CO_2 + N_2$	17.02% CO_2 + 82.98% N_2	Dalian Dart Gas Co., Dalain, China
Tetrabutylammonium bromide	TBAB	99% N_2	Shanghai McLean Biochemical Technology Co., Shanghai, China
Sodium dodecyl sulfate	SDS	92.5~99%	Shanghai McLean Biochemical Technology Co., Shanghai, China
Nano-aluminum oxide	Al_2O_3	20 wt% aqueous solution	Shanghai Aladdin Biochemical Technology Co., Shanghai, China
L-methionine	$C_5H_{11}NO_2S$	99%	Shanghai Titan Technology Co., Shanghai, China
L-leucine	$C_6H_{13}NO_2$	99%	Shanghai McLean Biochemical Technology Co., Shanghai, China

2.2.2. Experimental Methods

The specific experimental procedures are as follows:

(1) Water bath preparation: check and turn on the low-temperature thermostatic water bath, set the experimental temperature, and turn on the experimental data acquisition system and display terminal;
(2) Cleaning the reaction reactor: clean the reaction reactor with deionized water several times to avoid the residual experimental reagents affecting the accuracy of the experiment;
(3) Preparation of solution: use an electronic balance to accurately weigh the reagents, dissolve them in 150 mL deionized water, and make sure that they are fully stirred until they are evenly mixed;

(4) Solution injection: put the prepared solution as well as the rotor into the reactor, lock the reactor, and check the sealing;
(5) Pre-cooling: place the reaction reactor in a constant temperature water bath device, start the magnetic stirring system, and wait for the temperature in the reaction reactor to drop to a preset value and remain stable;
(6) Exhausting miscellaneous gases: connect the gas cylinder, control the pressure 1 MPa, purge the gas 3–4 times, and empty the original gas in the reactor;
(7) Injection of reaction gas: open the inlet road valves and the valve in front of the pressure gauge and start inflating the speed as far as possible to maintain uniformity to ensure that each group of experiments has the same injection time and ensure pressure stabilization and then close the inlet road valves and cylinders. Then, the reaction begins;
(8) Collect residual gas: after the reaction is finished, close the magnetic stirrer and collect residual gas;
(9) Collect decomposition gas: after collecting residual gas, release the gas in the reaction reactor and exhaust the original gas with a vacuum pump. Then, put it to room temperature in a stirring table to decompose, and collect decomposition gas when the room temperature is restored;
(10) Repeat operation steps (5)–(9), and perform two groups of the same experimental conditions;
(11) The end of the experiment: release the gas in the reaction reactor, dispose of the waste solution, and clean the reactor to prepare it for the next set of experiments;
(12) Use a gas chromatograph to detect the experimental gas.

The flowchart is presented in the accompanying Figure 2.

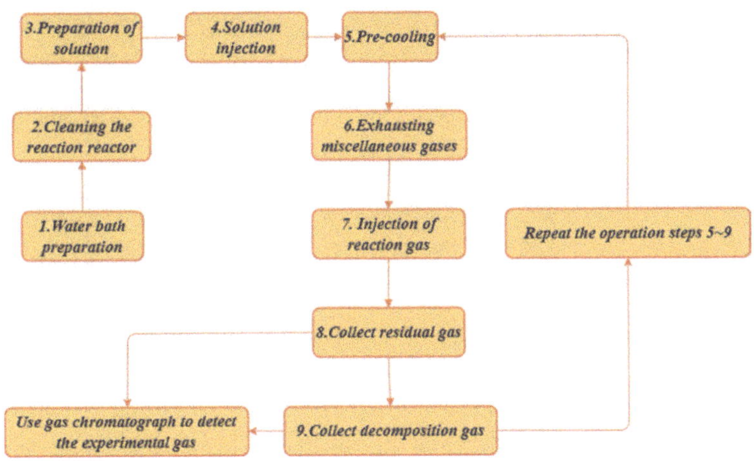

Figure 2. Experimental procedure.

The hydrate generation experiments were carried out under different working conditions with the application of promoters to study the effects of promoter type, promoter concentration, and initial pressure on the CO_2 hydrate generation characteristics and carbon capture capacity. In order to reduce the bias of experimental data and enhance the credibility of the results, it was ensured that each set of experiments was repeated at least twice. For Case A, the effects of a single thermodynamic promoter, TBAB concentration, and experimental pressure on the CO_2 hydrate generation process were investigated by fixing the temperature condition at 3 °C, and the TBAB concentration and experimental pressure were used as the conditioned variables for the experimental study. The experimental reagents were categorized into four concentrations of 3 wt%, 5 wt%, 7.5 wt%, and 10 wt%, and the pressures were 3 MPa and 4 MPa, respectively. For Case B, the effects of a single kinetic promoter concentration, as well as the experimental pressure on the CO_2 hydrate generation process, were explored due to the low CO_2 percentage of 17.02% in the simulated flue gas used for the experiments in this paper. Drawing on the experimental conditions studied by Linga et al. [48], the temperature of this experiment was set to two groups of 0 °C and 1 °C, and the initial pressure of the experiment was set to 10 MPa. The agents chosen were 0.1 wt% SDS [49], 0.3 wt% nano-Al_2O_3 [43], 0.1 wt% L-methionine [40] and 1 wt% L-leucine [50]. For Case C, the effect of the use of TBAB combined with the kinetic promoter on the CO_2 hydrate generation process was investigated; firstly, the optimal reaction temperature and pressure conditions and TBAB concentration were confirmed according to Case A, based on which the kinetic promoter was added to form a thermo–kinetic compound.

A total of 20 groups of experiments were carried out in the three cases, and the specific experimental conditions included experimental serial number, experimental temperature, experimental pressure, and experimental concentration, as shown in Table 2. One noteworthy aspect is the selection of temperature and pressure conditions throughout the phase equilibrium range for hydrate formation. In this study, we have chosen a temperature that is not exceedingly low while maintaining relatively mild pressure conditions. These findings have implications for subsequent industrial applications.

Table 2. Experimental conditions of CO_2 hydrate generation.

Experiment Number	Experimental Concentration/wt%	Initial Pressure/MPa	Experimental Temperature/°C
A1	5 wt% TBAB	3	3
A2	5 wt% TBAB	3	3
A3	5 wt% TBAB	4	3
A4	5 wt% TBAB	4	3
A5	7.5 wt% TBAB	3	3
A6	7.5 wt% TBAB	3	3
A7	7.5 wt% TBAB	4	3
A8	7.5 wt% TBAB	4	3
A9	10 wt% TBAB	3	3
A10	10 wt% TBAB	3	3
A11	10 wt% TBAB	4	3
A12	10 wt% TBAB	4	3
B1	0.1 wt% SDS	10	3
B2	0.3 wt% Nano Al_2O_3	10	3
B3	0.1 wt% L-methionine	10	3
B4	1 wt% L-leucine	10	3
C1	AS wt% TBAB + 0.1 wt% SDS	P_{AS}	T_{AS}
C2	AS wt% TBAB + 0.3 wt% Nano-Al_2O_3	P_{AS}	T_{AS}
C3	AS wt% TBAB + 0.1 wt% L-methionine	P_{AS}	T_{AS}
C4	AS wt% TBAB + 1 wt% L-leucine	P_{AS}	T_{AS}

2.3. Experimental Data Processing

2.3.1. Induction Time

As shown in Figure 3, the hydrate formation process involves three key steps: first, the dissolution of gas, then nucleation, and finally, the growth of crystal nuclei [51]. The induction period is a period of time before the hydrate begins to form, and with the gradual formation of the hydrate, a large amount of heat will be released, leading to a sharp rise in temperature and the formation of an obvious exothermic peak, which is used to determine the generation and induction time of the hydrate.

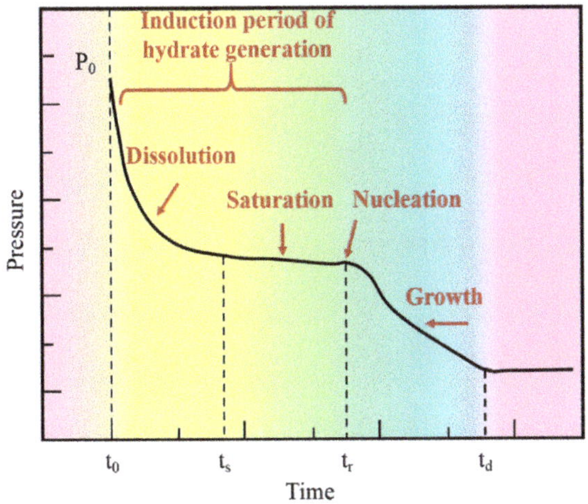

Figure 3. The formation of hydrates is described as occurring through three distinct phases: saturation, nucleation, and growth.

2.3.2. Calculation of Gas Separation Coefficient

The gas separation coefficient is a physical quantity that describes the relative degree of separation of gases under specific conditions at a certain temperature and pressure. The CO_2 gas separation coefficient χ in this experimental system can be calculated from the molar ratio of N_2 and CO_2 in the gas phase before and after the reaction.

$$\chi = \frac{y_{N_2,g}/y_{CO_2,g}}{y_{N_2,f}/y_{CO_2,f}} \tag{1}$$

$y_{CO_2,f}$ and $y_{CO_2,g}$ are the molar ratios of CO_2 in the gas phase at the beginning and the end of the reaction, respectively; $y_{N_2,f}$ and $y_{N_2,g}$ are the molar ratios of N_2 in the gas phase at the beginning and the end of the reaction.

2.3.3. Calculation of Gas Consumption

The ideal Equation of state for CO_2 gas is as follows:

$$P_i V = nZRT_i \tag{2}$$

Pi and *Ti* are the pressure and temperature of the gas in the reaction reactor fed back by the temperature and pressure acquisition equipment at i time, respectively;

V is the volume of the gas phase in the reaction reactor, $V = 2 \times 10^{-7}$ m^3;

R represents the ideal gas constant, R = 8.314 kJ/(kmol·K);

Zi represents the gas compression factor at moment i, which can be calculated from Pitzer's Equation (3):

$$z_i = 1 + \left[0.083 - \frac{0.422}{T_r^{1.6}}\right] \times \frac{P_r}{T_r} + w \times \left[0.139 - \frac{0.172}{T_r^{4.2}}\right] \times \frac{P_r}{T_r} \tag{3}$$

w represents the eccentricity factor of CO_2, $w = 0.022394$;

Pr and *Tr* represent the contrast pressure and contrast temperature of the gas, respectively, which can be calculated according to Equations (4) and (5):

$$P_r = \frac{P_i}{P_c} \tag{4}$$

$$T_r = \frac{T_i}{T_c} \tag{5}$$

In Equations (4) and (5), *Pc* represents the critical pressure of CO_2 $Pc = 7.45$ MPa; *Tc* represents the critical temperature of CO_2 $Tc = 304.25K$.

According to Equation (6), the molar amount of gas consumed Δn at t time for CO_2 hydrate generation can be calculated:

$$\Delta n = n_0 - n_t = \frac{V}{R}\left(\frac{P_0}{Z_0 T_0} - \frac{P_t}{Z_t T_t}\right) \tag{6}$$

Δn denotes the gas consumption, mol; V denotes the volume of the gas phase in the reactor, m^3;

P_0, T_0 and P_t, T_t are the pressure and temperature of the initial and final states, respectively;

R denotes the molar gas constant, R = 8.3145 J/(mol·K);

Z denotes the gas compression factor.

3. Results and Discussion

3.1. Influence of Thermodynamic Additives on CO_2 Hydrate Generation Characteristics

Excessively high phase equilibrium conditions for CO_2 hydrate generation within associated gas and industrial waste gas is a key factor limiting the commercial application of CO_2 capture by the hydrate method. Thermodynamic promoters have shown significant effects in reducing the hydrate phase equilibrium. Therefore, in this paper, the most representative thermodynamic additive TBAB was selected to carry out a series of experiments and coupled with the effect of pressure to investigate the influence of TBAB concentration and pressure on the characteristics of CO_2 hydrate generation in the flue gas, which provides theoretical guidance for efficient CO_2 capture.

3.1.1. Influence of TBAB on CO_2 Hydrate Generation Characteristics

As shown in Figure 4, under the experimental conditions of Case A, when 5 wt% TBAB was added to the flue gas system, the induction times of CO_2 hydrate generation were 75 and 41 min, while when the TBAB concentration was increased to 7.5 wt%, the induction time are 51 and 37 min. When the TBAB concentration was further increased to 10 wt%, the hydrate generation induction time decreased to 4 and 23 min. A negative correlation was observed between the experimental concentration and the induction time (Figure 5). This phenomenon suggests that the cationic TBA+ is able to promote the formation of hydrate nucleosomes with the increase of TBAB concentration, and thus, the hydrate generation induction time is gradually shortened with the increase of TBAB concentration [52]. In addition, the differences in the hydrate generation induction time obtained in two sets of replicate experiments with the same concentration of TBAB suggest a high degree of stochasticity in the hydrate reaction. At the microscopic level, this is mainly due to the instability of the hydrate nucleation.

After the formation of CO_2 hydrate, the average gas consumption in the three concentration systems of 5 wt%, 7.5 wt%, and 10 wt% is summarized in Table 3. The gas consumption values of the 5.0 wt% and 7.5 wt% systems are close to 0.0225 mol and 0.0222 mol, respectively, while the 10.0 wt% system has the smallest gas consumption of 0.0119 mol, which is only half of the values of the 5.0 wt% and 7.5 wt% systems. Considering the observations of the experimental process, it is hypothesized that at some point in the experiment, a magnetic stirrer obstruction event may occur in the 10.0 wt% experimental group, and due to the fast rate of hydrate generation and high concentration in the early stage of the experiment, the magnetic stirrer was subjected to increase resistance, which resulted in a slowing down of the rate of hydrate generation and a decrease in gas consumption in the later stage of the experiment. Combining the induction time, gas consumption, and economic cost, 5 wt% TBAB concentration is the best choice for promoting CO_2 capture based on the no-hydrate method in this study system.

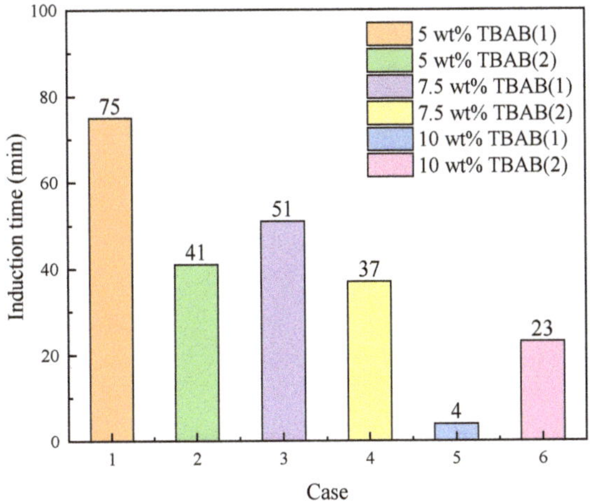

Figure 4. Induction time of CO_2 hydrate generation in TBAB system with different concentrations at 4 MPa.

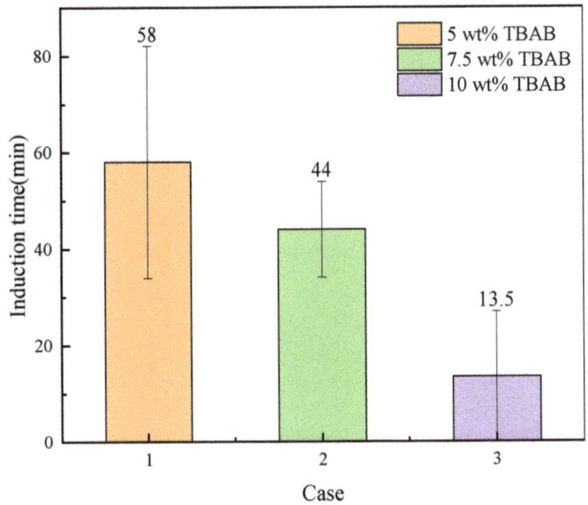

Figure 5. Error bars for induction time of CO_2 hydrate generation at 4 MPa for TBAB systems with different concentrations.

Table 3. Gas consumption in hydrate formation experiments in the system with different concentrations of TBAB at an initial pressure of 4 MPa.

Initial Conditions	TBAB Concentration	Final Gas Consumption
3 °C 4 MPa	5 wt% 7.5 wt% 10 wt% 0.0119 mol	0.0225 mol 0.0222 mol 0.0119 mol

3.1.2. Effect of Pressure on CO_2 Hydrate Generation Characteristics

Pressure is a key factor affecting hydrate generation and determining production safety. Therefore, in order to reveal the specific effect of pressure on the hydrate generation rate, the induction time and gas consumption of hydrate generation were experimentally investigated under two pressure conditions, 3 MPa, and 4 MPa. Figure 6 shows the relationship between the hydrate reaction pressure and the average induction time, the induction time shortening as the pressure increases. In the 5 wt% TBAB system, the induction time of hydrate generation was 78 min for the experimental pressure of 3 MPa, which was reduced to 53 min when the experimental pressure was increased to 4 MPa, a decrease of 32.05%; in the 7.5 wt% TBAB system, the induction time of hydrate generation was 75 min for the experimental pressure of 3 MPa, which was reduced to 53 min when the experimental pressure was increased to 4 MPa. the induction time was reduced to 44 min, with a decrease of 41.33%; this phenomenon indicates that under the condition of the same reaction temperature, the increase of the hydrate reaction pressure will promote the gas escape, increase the gas–liquid contact area, and then accelerate the hydrate nucleation rate, which ultimately leads to the shortening of the induction period of hydrate [53].

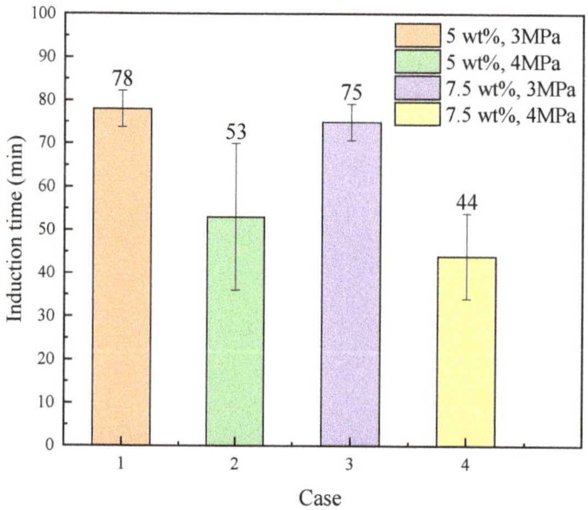

Figure 6. Different pressures and TBAB concentrations promote hydrate generation induction time graphs.

As shown in Figure 7, in the process of gas hydrate formation, for the experimental system with the same TBAB concentration, the gas consumption rate and final gas consumption increase with the reaction pressure increase. Among them, in the system with a pressure of 4 MPa and a TBAB concentration of 7.5 wt%, the phenomenon of secondary hydrate formation also occurred. This is because, as the initial experimental pressure increases, the gas supersaturation in the system increases, and the hydrate growth drive force increases, which in turn leads to an increase in the hydrate growth rate and final gas consumption [53]. This suggests that high pressure can accelerate the hydrate formation process and effectively capture CO_2 from the industrial waste gas within the safe production range.

3.1.3. Synergistic Effect of TBAB Concentration and Pressure on CO_2 Hydrate Formation Characteristics

Table 4 shows the effects of TBAB concentration and reaction pressure on the CO_2 gas separation coefficient: with the increase of TBAB concentration, the CO_2 gas separation coefficient decreases; with the increase of reaction pressure, the CO_2 gas separation coefficient increases. This is consistent with the experimental data on the effect of gas consumption in Section 3.3.2, and the CO_2 gas separation coefficients of the [12] groups of experiments are all greater than 1, which indicates that TBAB can separate CO_2 by the hydrate method. The experimental results of control experimental groups with different pressures corresponding to each concentration are summarized in Figure 8, from which it can be seen that the average induction time, gas consumption, and separation coefficient of the 4 MPa experimental group are better than those of the 3 MPa experimental group, except that the separation coefficient of the 4 MPa experimental group is slightly smaller than that of the 3 MPa experimental group in the 10 wt% experimental group, which is presumed due to the weakening effect of the concentration on the separation coefficient than the promoting effect of the pressure when the concentration of TBAB is too high. The weakening effect of concentration is greater than the promoting effect of pressure.

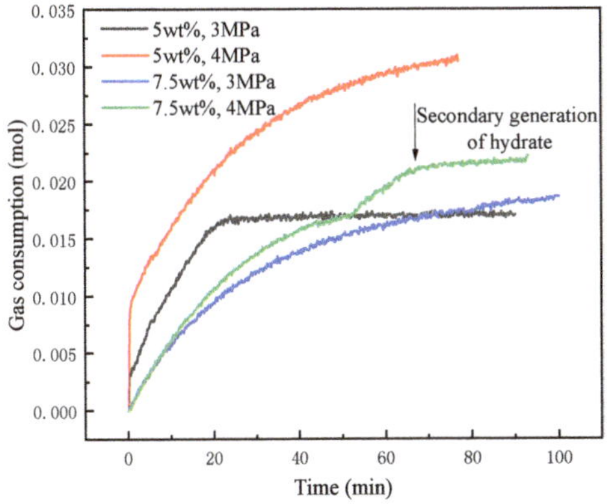

Figure 7. Curve of gas consumption versus time for hydrate formation experiments with different pressures and TBAB concentrations.

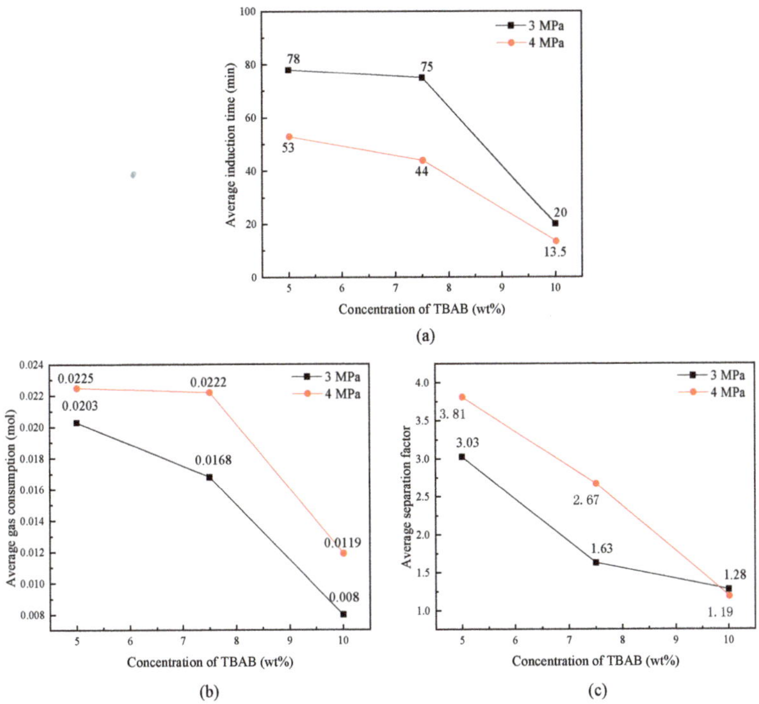

Figure 8. Comparison of experimental results of different pressures and TBAB concentrations during the hydrate formation process. (**a**) The average induction time of hydrate formation under different pressure and concentration of TBAB. (**b**) Average gas consumption of hydrate formation under different pressure and concentration of TBAB. (**c**) Average separation coefficient under different pressure and concentration of TBAB.

Table 4. Experimental data of CO_2 hydrate generation at 3 °C.

Experiment No.	Initial Pressure (MPa)	TBAB Concentration (wt%)	Molar Proportion of CO_2 in the Gas Phase at the End of the Reaction (%)	Molar Proportion of N_2 in the Gas Phase at the End of the Reaction (%)	CO_2 Gas Separation Coefficient	Average Value of Separation Coefficient
1	3	5	6.22	93.78	3.10	3.03
2	3	5	6.51	93.49	2.96	
3	3	7.5	11.08	88.92	1.65	1.63
4	3	7.5	11.34	88.66	1.61	
5	3	10	14.1	85.9	1.25	1.28
6	3	10	13.6	86.4	1.31	
7	4	5	3.39	94.33	3.41	3.81
8	4	5	4.64	95.36	4.23	
9	4	7.5	7.01	92.99	2.73	2.67
10	4	7.5	7.32	92.68	2.61	
11	4	10	15.36	84.63	1.13	1.19
12	4	10	14.24	85.75	1.24	

Overall consideration, when the temperature is 3 °C, the 5 wt% TBAB concentration is more suitable for the separation of CO_2 within the industrial waste gas by using the hydrate method, and the best CO_2 separation effect is achieved when the pressure is 4 MPa, with a separation factor of 3.81.

3.2. Effect of Kinetics Additives on the Characteristics of CO_2 Hydrate Formation in the Industrial Waste Gas System

Unlike thermodynamic catalysts, kinetic promoters mainly increase the hydrate generation rate without participating in the hydrate [54] formation process [54]. Under the condition of Case C, four single kinetic promoters were selected for testing experiments, mainly analyzing the induction period of CO_2 hydrate formation and CO_2 separation coefficients with the 0.1 wt% SDS addition; the results are shown in Figure 8.

From Figure 9, it can be observed that during the whole reaction process, the temperature in the reactor only fluctuated within a certain small range, the temperature curve did not appear to have obvious exothermic peaks, and through the visual window of the reactor, we did not observe the formation of hydrate during the experiment, and we subsequently repeated the experiment many times under the same conditions, which also did not generate hydrate. The residual gas was collected after the pressure in the reactor no longer changed, and the residual gas components were analyzed by gas chromatography. The CO_2 gas separation coefficient was 2.19 when the reaction temperature was 1 °C and 2.80 when the reaction temperature was 0 °C. Although the value of the gas separation coefficient was greater than 1, indicating a certain carbon trapping ability in this set of experiments, the pressure (10 MPa) in this set of experiments was much greater than the reaction pressures (3 MPa and 4 MPa) in Case A, thus increasing the CO_2 solubility to a certain extent, but it was still small compared with the CO_2 separation coefficient in Case A. The CO_2 separation coefficient in this set of experiments was also greater than that in Case A. The CO_2 separation coefficient in this set of experiments was greater than that in Case A. Subsequent test experiments with the addition of single nano-Al_2O_3, L-methionine, and L-leucine were also conducted, all of which also failed to produce hydrates and had poor separation factors. The above results show that the formation of CO_2 hydrate can enhance the CO_2 separation and capture effect to a greater extent; pressurization and the addition of surfactants can only increase the solubility of CO_2 in water to a certain extent, which not only makes the reaction conditions more severe but also makes the CO_2 capture effect not very satisfactory. Therefore, this paper concludes that the hydrate method of capturing low-concentration CO_2 in industrial waste gas needs to be coupled with thermal-kinetic promoters to achieve a better CO_2 capture effect.

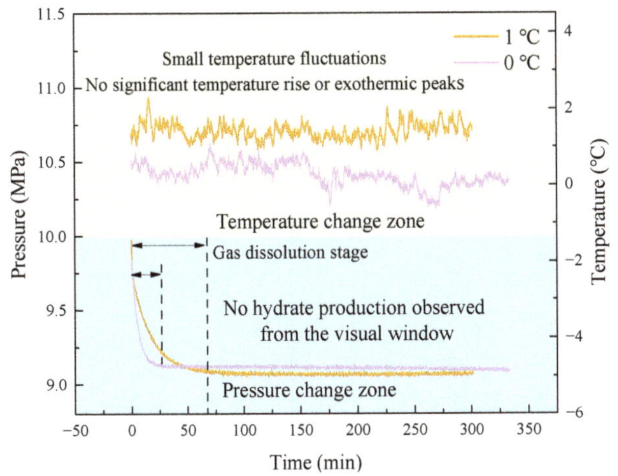

Figure 9. Temperature and pressure change curve of CO_2 hydrate formation process in 0.1 wt% SDS solution.

3.3. *Experimental Study of CO_2 Hydrate Formation Characteristics in Thermal/Kinetic Combination Promoter System*

In order to investigate the effect of TBAB coupled with the kinetic promoter on the CO_2 hydrate formation characteristics, firstly, according to the experimental results in Section 3.1, the concentration of TBAB with optimal CO_2 capture effect under the conditions (4 MPa, 3 °C) was determined to be 5 wt%, and with reference to the concentration of kinetic promoter in other pieces of literature, four groups of compound promoter concentrations were set up as 5 wt%, respectively. TBAB + 0.1 wt% SDS, 5 wt% TBAB + 0.3 wt% nano-Al_2O_3, 5 wt% TBAB + 0.1 wt% L-methionine, 5 wt% TBAB + 1 wt% L-leucine, to carry out experimental studies on the induction time, gas consumption, and separation factor of the generation of CO_2 hydrate.

3.3.1. Effect of Different Compounding Accelerators on the Induction Time of CO_2 Hydrate Formation

The induction time of CO_2 hydrate formation in five groups is shown in Figure 10. The addition of 0.1 wt% SDS, 0.3 wt% nano-Al_2O_3, 0.1 wt% L-methionine, and 1 wt% L-leucine to 5 wt% TBAB reduced the induction time of CO_2 hydrate formation to 37 min, 21 min, 4 min, and 3 min, which were 31.5%, 61.1%, 92.6%, and 94.6%, respectively, compared with the addition of a single 5 wt% TBAB system. In particular, the reduction in the induction time of the experimental system with SDS and nano-Al_2O_3 was smaller and less facilitated than that of the experimental systems with L-methionine and L-leucine. In addition, the error columns of the induction time of the three experimental groups of a single 5 wt% TBAB and the addition of 0.1 wt% SDS and 0.3 wt% nano-Al_2O_3 are large, which indicate that the reaction stability of these three complex combinations was poor. Therefore, from the analysis of the results of this experiment, it can be concluded that the combination of TBAB with different kinetic promoters can effectively reduce the induction period of hydrate generation, and the system with the addition of 0.1 wt% L-methionine and 1 wt% L-leucine has the shortest induction time and high reaction stability. It is noteworthy that TBAB is a quaternary ammonium salt, wherein the cation is capable of forming a half-cage structure with water molecules. Additionally, L-methionine and leucine, as amino acids, contain amino and carboxyl groups that can form hydrogen bonds with water molecules, thereby enhancing the interactions between water molecules and thus contributing to the formation of hydrates.

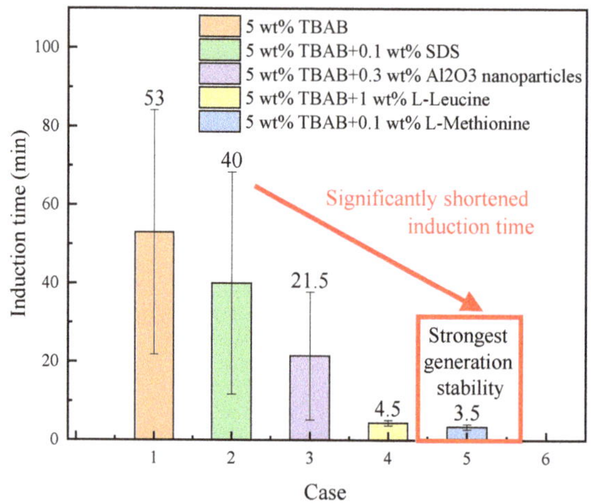

Figure 10. Induction time of CO_2 hydrate generation by TBAB combined with different kinetic promoters.

3.3.2. Effect of Different Compounding Promoters on Gas Consumption for CO_2 Hydrate Formation

The total gas consumption data of CO_2 hydrate formation in different complexed promoter systems are presented in Figure 11, and it was found that the addition of 0.1 wt% SDS was effective in increasing the final gas consumption, while the addition of 0.3 wt% nano-Al_2O_3, 0.1 wt% L-methionine, and 1 wt% L-leucine enhanced the final gas consumption by 42~45%. Zhang [55] et al. proposed that amino acids can be used to increase CO_2 hydrate gas storage capacity by capillary action surface adsorption behavior, and the CO_2 hydrate generated in the L-methionine system has a more porous structure, which can provide more space for gas–liquid contact. Al_2O_3 nanoparticles have a large specific surface area, which can effectively enhance the mass transfer capacity of the solution and thus promote CO_2 dissolution, significantly shorten the induction period of hydrate, and increase the gas consumption; in addition, Al_2O_3 also improves the thermal conductivity of the solution, which helps to transfer the heat released from the hydrate formation, and further promotes the hydrate generation [56]. Therefore, from the analysis of the experimental results, it can be concluded that the combined kinetic promoter of TBAB can effectively improve the gas consumption of CO_2 capture by the hydrate method, among which the system with the addition of 0.3 wt% nano-Al_2O_3, 0.1 wt% L-methionine, and 1 wt% L-leucine showed the best CO_2 capture effect and high reaction stability.

It should be noted, however, that nano-Al_2O_3 can be prepared by a variety of methods, including laser-induced vapor deposition, plasma vapor synthesis, chemical vapor deposition (CVD), and solid-phase synthesis (e.g., these include mechanical comminution, amorphous crystallization, pyrolysis, combustion, and liquid-phase methods such as precipitation, sol–gel, hydrothermal synthesis, microemulsion, and electrochemistry). The utilization of nano-Al_2O_3 may potentially give rise to adverse effects, including the generation of harmful gases during the preparation process, the transportation, and transformation of nanoparticles with a high specific surface area and activity in the environment, and the inhalation of nano-Al_2O_3 dust, which could have detrimental effects on human health.

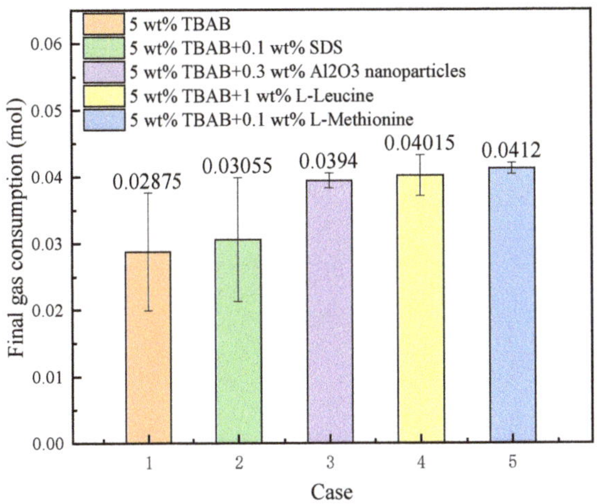

Figure 11. Schematic diagram of total gas consumption of CO_2 hydrate by different kinetic promoters combined with TBAB.

3.3.3. Effect of Different Compounding Promoters on the Separation Factor of CO_2 Hydrate Formation

Figure 12 shows the values and errors of the CO_2 gas separation coefficient in different kinetic promoters combined with the TBAB system. When different kinetic promoters were added to the 5 wt% TBAB solution, the separation coefficient of CO_2 decreased slightly. It was hypothesized that it might be possible that the surfactant SDS could reduce the surface tension of water, the nano-Al_2O_3 had a high specific surface area, and both L-methionine and L-leucine belonged to nonpolar amino acids, and the nature of these solutions enhanced the solubility of CO_2 [57], and some of the CO_2 gases were left in the solution at the end of the hydrate decomposition process, which resulted in a slight decrease in the CO_2 separation coefficient.

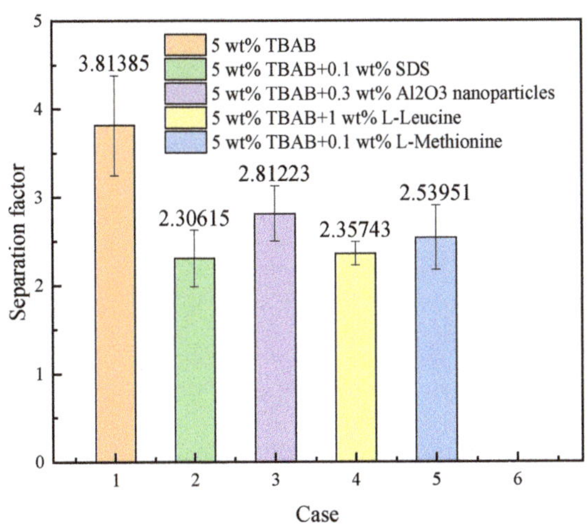

Figure 12. Schematic representation of CO_2 hydrate separation coefficient in the TBAB system with different kinetic promoters.

4. Conclusions and Prospect

4.1. Conclusions

In this paper, we take the CO_2 capture and efficiency enhancement in the industrial waste gas of hydrate method as the research objective and carry out experimental research around the optimization of hydrate generation promoter. First, we optimized the concentration of the thermodynamic additive TBAB and the reaction pressure, and then we carried out the rapid generation of CO_2 hydrate under the conditions of different TBAB-kinetic promoter combinations and studied the effect of the combinations of promoter on the characteristics of the generation of CO_2 hydrate in the industrial waste gas. The effect of the compounding promoters on the characteristics of CO_2 hydrate generation in the industrial waste gas was investigated. The specific conclusions are as follows:

(1) The TBAB concentration of 5 wt%, with the largest gas consumption (0.0225 mol) and separation coefficient (3.81), and the relatively short average induction time, is the optimal choice for the industrial waste gas to promote CO_2 hydrate formation in this paper. It provides a theoretical basis for the concentration of high efficiency for subsequent industrial exhaust gas treatment;

(2) The average induction time, gas consumption, and separation coefficient of the 4 MPa experimental group are better than those of the 3 MPa experimental group, but it should be specified that the higher reaction pressure would increase the trapping cost and decrease the safety. The complementary effect of the higher-pressure environment is better than that of the lower-pressure environment, and the ensuing safety issue is worth considering for the subsequent industrial waste gas treatment;

(3) In this paper, the TBAB system is studied in combination with a kinetic promoter. Compared with the pure TBAB system, the introduction of kinetic promoters significantly shortened the induction time and increased the gas consumption of CO_2 hydrate. Among them, the shortest induction time was 3.5 min in the L-methionine-containing experimental group with the strongest stability of hydrate formation, and the highest gas consumption was observed in the L-methionine-containing nano group and the L-leucine-containing experimental group with the gas consumption of 0.0412 mol and 0.04015 mol, respectively; however, the incorporation of kinetic promoters slightly decreased the gas separation coefficient of CO_2 compared with that of the pure TBAB system. In summary, 5 wt% TBAB + 0.1 wt% L-methionine is the best promoter compounding scheme in this paper, which has the shortest induction time, the largest gas consumption, and the best separation coefficient.

4.2. Outlook

In order to promote the further practical application of hydrate technology, it is necessary to carry out more in-depth research work, which can be carried out in the following aspects:

(1) The additives used in the experiments in this paper are conventional types in the laboratory, and the carbon capture effect has the prospect of further improvement, which should continue to carry out the screening, research, and development of new additives;

(2) Follow-up experiments can also be changed by using the solution spray method, bubble method, and other ways to increase the gas–liquid contact area to improve the effect of gas separation further;

(3) Further study of the effect of concentration ratios and multivariate combinations of thermodynamic compounding promoters on the CO_2 hydrate formation characteristics of industrial waste gas.

Author Contributions: Writing—review and editing, Y.G.; writing—original draft preparation, H.C.; formal analysi, R.Q.; data curation, H.Y.; Conceptualization, T.H.; validation, X.L.; project administration and supervision, H.L. and S.S. All authors have read and agreed to the published version of the manuscript.

Funding: This research was funded by [the State Key Laboratory of Offshore Natural Gas Hydrates] grant number [2022-KFJJ-SHW], [the Zhejiang Provincial Natural Science Foundation] grant number [No. LQ24E060008], [the Natural Science Foundation of Ningbo] grant number [2022J005], and [the Beijing New-star Plan of Science and Technology] grant number [NO.20240484721].

Data Availability Statement: Data are contained within the article.

Conflicts of Interest: Authors Yang Ge, Haihong Chen, Rui Qin, Haiyuan Yao and Ting Huang were employed by the company China National Offshore Oil Corporation Research Institute Co., Ltd. The remaining authors declare that the research was conducted in the absence of any commercial or financial relationships that could be construed as a potential conflict of interest.

References

1. Liu, Z.; Deng, Z.; Davis, S.J.; Ciais, P. Global carbon emissions in 2023. *Nat. Rev. Earth Environ.* **2024**, *5*, 253–254. [CrossRef]
2. Ritchie, H.; Rosado, P.; Roser, M. *CO_2 and Greenhouse Gas Emissions*; OurWorldInData.org: Oxford, UK, 2023.
3. Thippeshnaik, G.; Prakash, S.B.; Suresh, A.B.; Chandrashekarappa, M.P.G.; Samuel, O.D.; Der, O.; Ercetin, A. Experimental Investigation of compression Ignition Engine combustion, Performance, and Emission Characteristics of Ternary Blends with Higher Alcohols (1-Heptanol and n-Octanol). *Energies* **2023**, *16*, 6582. [CrossRef]
4. Godwin, D.J.; Varuvel, E.G.; Martin, M.L.J. Prediction of combustion, performance, and emission parameters of ethanol powered spark ignition engine using ensemble Least Squares boosting machine learning algorithms. *J. Clean. Prod.* **2023**, *421*, 138401. [CrossRef]
5. Budzianowski, W.M. Implementing carbon capture, utilisation and storage in the circular economy. *Int. J. Glob. Warm.* **2017**, *12*, 272–296. [CrossRef]
6. Mohsin, I.; Al-Attas, T.A.; Sumon, K.Z.; Bergerson, J.; Mccoy, S.; Kibria, M.G. Economic and Environmental Assessment of Integrated Carbon Capture and Utilization. *Cell Rep. Phys. Sci.* **2020**, *1*, 100104. [CrossRef]
7. Ishaq, H.; Ali, U.; Sher, F.; Anus, M.; Imran, M. Process analysis of improved process modifications for ammonia-based post-combustion CO_2 capture. *J. Environ. Chem. Eng.* **2021**, *9*, 104928. [CrossRef]
8. Sneddon, G.; Greenaway, A.; Yiu, H.H.P. The Potential Applications of Nanoporous Materials for the Adsorption, Separation, and Catalytic Conversion of Carbon Dioxide. *Adv. Energy Mater.* **2014**, *4*, 1301873. [CrossRef]
9. Shen, M.; Tong, L.; Yin, S.; Liu, C.; Wang, L.; Feng, W.; Ding, Y. Cryogenic technology progress for CO_2 capture under carbon neutrality goals: A review. *Sep. Purif. Technol.* **2022**, *299*, 121734. [CrossRef]
10. Fu, L.; Ren, Z.; Si, W.; Ma, Q.; Huang, W.; Liao, K.; Huang, Z.; Wang, Y.; Li, J.; Xu, P. Research progress on CO_2 capture and utilization technology. *J. CO2 Util.* **2022**, *66*, 102260. [CrossRef]
11. Hou, R.; Fong, C.; Freeman, B.D.; Hill, M.R.; Xie, Z. Current status and advances in membrane technology for carbon capture. *Sep. Purif. Technol.* **2022**, *300*, 121863. [CrossRef]
12. Raganati, F.; Miccio, F.; Ammendola, P. Adsorption of Carbon Dioxide for Post-Combustion Capture: A Review. *Energy Fuels* **2021**, *35*, 12845–12868. [CrossRef]
13. Xu, C.-G.; Chen, Z.-Y.; Cai, J.; Li, X.-S. Study on Pilot-Scale CO_2 Separation from Flue Gas by the Hydrate Method. *Energy Fuels* **2014**, *28*, 1242–1248. [CrossRef]
14. Ricaurte, M.; Torré, J.-P.; Asbai, A.; Broseta, D.; Dicharry, C. Experimental Data, Modeling, and Correlation of Carbon Dioxide Solubility in Aqueous Solutions Containing Low Concentrations of Clathrate Hydrate Promoters: Application to CO_2–CH_4 Gas Mixtures. *Ind. Eng. Chem. Res.* **2012**, *51*, 3157–3169. [CrossRef]
15. Sun, H.R.; Chen, J.; Ji, X.; Karunakaran, G.; Chen, B.B.; Ranjith, P.G.; Song, Y.C.; Yang, M.J. Optimizing CO_2 hydrate storage: Dynamics and stability of hydrate caps in submarine sediments. *Appl. Energy* **2024**, *376*, 124309. [CrossRef]
16. Aminnaji, M.; Qureshi, M.F.; Dashti, H.; Hase, A.; Mosalanejad, A.; Jahanbakhsh, A.; Babaei, M.; Amiri, A.; Maroto-Valer, M. CO_2 Gas hydrate for carbon capture and storage applications—Part 1. *Energy* **2024**, *300*, 131579. [CrossRef]
17. Hu, Q.; Xiao, X. Formation methods and applications of carbon dioxide hydrate: An overview. *Carbon Capture Sci. Technol.* **2023**, *7*, 100113. [CrossRef]
18. Ge, Y.; Wang, L.; Feng, K.; Shen, S.; Liu, Y.; Wu, Z.; Liu, Z.; Song, Y. High-pressure and temperature-controlled CPT calibration chamber based on deep-sea hydrate reservoirs. *Geoenergy Sci. Eng.* **2024**, *242*, 213207. [CrossRef]
19. Liu, W.; Liu, H.; Wang, L.; Feng, K.; Shen, S.; Liu, Y.; Zhao, J.; Li, Y. Undrained Mechanical Properties of Overlying Soils of Gas Hydrate Reservoirs in the South China Sea. *Energy Fuels* **2024**, *38*, 3020–3031. [CrossRef]
20. Wang, L.; Shen, S.; Wu, Z.; Wu, D.; Li, Y. Strength and creep characteristics of methane hydrate-bearing clayey silts of the South China Sea. *Energy* **2024**, *294*, 130789. [CrossRef]
21. Wang, L.; Zhao, J.; Sun, X.; Wu, P.; Shen, S.; Liu, T.; Li, Y. Comprehensive review of geomechanical constitutive models of gas hydrate-bearing sediments. *J. Nat. Gas Sci. Eng.* **2021**, *88*, 103755. [CrossRef]
22. Zhang, L.; Wang, Y.; Lang, C.; Yang, L.; Zhao, J.; Song, Y. Crystal Growth of CO_2–CH_4 Hydrate on a Solid Surface with Varying Wettability in the Presence of PVCap. *Cryst. Growth Des.* **2024**, *24*, 4697–4706. [CrossRef]
23. Chen, B.; Li, K.; Sun, H.; Jiang, L.; Yang, M.; Song, Y. Promoting Effect of Common Marine Cations on Hydrate Dissociation and Structural Evolution under a Static Electric Field. *J. Phys. Chem. B* **2023**, *127*, 698–709. [CrossRef] [PubMed]

24. Linga, P.; Kumar, R.; Englezos, P. The clathrate hydrate process for post and pre-combustion capture of carbon dioxide. *J. Hazard. Mater.* **2007**, *149*, 625–629. [CrossRef] [PubMed]
25. Nguyen, N.N.; La, V.T.; Huynh, C.D.; Nguyen, A.V. Technical and economic perspectives of hydrate-based carbon dioxide capture. *Appl. Energy* **2022**, *307*, 118237. [CrossRef]
26. Wang, L.; Sun, X.; Shen, S.; Wu, P.; Liu, T.; Liu, W.; Zhao, J.; Li, Y. Undrained triaxial tests on water-saturated methane hydrate–bearing clayey-silty sediments of the South China Sea. *Can. Geotech. J.* **2021**, *58*, 351–366. [CrossRef]
27. Wang, L.; Li, Y.; Shen, S.; Liu, W.; Sun, X.; Liu, Y.; Zhao, J. Mechanical behaviours of gas-hydrate-bearing clayey sediments of the South China Sea. *Environ. Geotech.* **2022**, *9*, 210–222. [CrossRef]
28. Chen, B.; Sun, H.; Li, K.; Yu, T.; Jiang, L.; Yang, M.; Song, Y. Unsaturated water flow-induced the structure variation of gas hydrate reservoir and its effect on fluid migration and gas production. *Energy* **2023**, *282*, 128843. [CrossRef]
29. Wang, X.; Zhang, Y.; Wang, F.; Yin, Z.; Zhang, Z.; Ting, V.P. Unlocking the potential of hydrate-based carbon capture: A review of passive techniques for CO_2 hydrate formation promotion. *Gas Sci. Eng.* **2024**, *126*, 205323. [CrossRef]
30. Ma, S.; Liu, Z.; Zheng, J.-n.; Wu, Z.; Li, N.; Guan, X.; Han, J.-n.; Yang, M.; Song, Y. Natural Gas Hydrate Decomposition Characteristics at the Exploitation Anaphase via Sediment Warming. *Energy Fuels* **2024**, *38*, 14334–14342. [CrossRef]
31. Li, Z.; Zhong, D.-L.; Zheng, W.-Y.; Yan, J.; Lu, Y.-Y.; Yi, D.-T. Morphology and kinetic investigation of TBAB/TBPB semiclathrate hydrates formed with a CO_2 + CH_4 gas mixture. *J. Cryst. Growth* **2019**, *511*, 79–88. [CrossRef]
32. Li, Y.; Gambelli, A.M.; Rossi, F.; Mei, S. Effect of promoters on CO_2 hydrate formation: Thermodynamic assessment and microscale Raman spectroscopy/hydrate crystal morphology characterization analysis. *Fluid Phase Equilibria* **2021**, *550*, 113218. [CrossRef]
33. Li, S.; Fan, S.; Wang, J.; Lang, X.; Liang, D. CO_2 capture from binary mixture via forming hydrate with the help of tetra-n-butyl ammonium bromide. *J. Nat. Gas Chem.* **2009**, *18*, 15–20. [CrossRef]
34. Duc, N.H.; Chauvy, F.; Herri, J.-M. CO_2 capture by hydrate crystallization—A potential solution for gas emission of steelmaking industry. *Energy Convers. Manag.* **2006**, *48*, 1313–1322. [CrossRef]
35. Tang, J.; Zeng, D.; Wang, C.; Chen, Y.; He, L.; Cai, N. Study on the influence of SDS and THF on hydrate-based gas separation performance. *Chem. Eng. Res. Des.* **2013**, *91*, 1777–1782. [CrossRef]
36. Kim, S.; Seo, Y. Semiclathrate-based CO_2 capture from flue gas mixtures: An experimental approach with thermodynamic and Raman spectroscopic analyses. *Appl. Energy* **2015**, *154*, 987–994. [CrossRef]
37. Liu, F.-P.; Li, A.-R.; Qing, S.-L.; Luo, Z.-D.; Ma, Y.-L. Formation kinetics, mechanism of CO_2 hydrate and its applications. *Renew. Sustain. Energy Rev.* **2022**, *159*, 112221. [CrossRef]
38. Zhou, S.-D.; Yu, Y.-S.; Zhao, M.-M.; Wang, S.-L.; Zhang, G.-Z. Effect of Graphite Nanoparticles on Promoting CO_2 Hydrate Formation. *Energy Fuels* **2014**, *28*, 4694–4698. [CrossRef]
39. Hassan, H.; Javidani, A.M.; Mohammadi, A.; Pahlavanzadeh, H.; Abedi-Farizhendi, S.; Mohammadi, A.H. Effects of Graphene Oxide Nanosheets and Al_2O_3 Nanoparticles on CO_2 Uptake in Semi-clathrate Hydrates. *Chem. Eng. Technol.* **2021**, *44*, 48–57. [CrossRef]
40. Liu, X.; Ren, J.; Chen, D.; Yin, Z. Comparison of SDS and L-Methionine in promoting CO_2 hydrate kinetics: Implication for hydrate-based CO_2 storage. *Chem. Eng. J.* **2022**, *438*, 135504. [CrossRef]
41. Yi, J.; Zhong, D.-L.; Yan, J.; Lu, Y.-Y. Impacts of the surfactant sulfonated lignin on hydrate based CO_2 capture from a CO_2/CH_4 gas mixture. *Energy* **2019**, *171*, 61–68. [CrossRef]
42. Zhong, Y.; Rogers, R.E. Surfactant effects on gas hydrate formation. *Chem. Eng. Sci.* **2000**, *55*, 4175–4187. [CrossRef]
43. Said, S.; Govindaraj, V.; Herri, J.-M.; Ouabbas, Y.; Khodja, M.; Belloum, M.; Sangwai, J.S.; Nagarajan, R. A study on the influence of nanofluids on gas hydrate formation kinetics and their potential: Application to the CO_2 capture process. *J. Nat. Gas Sci. Eng.* **2016**, *32*, 95–108. [CrossRef]
44. Li, Y.; Gambelli, A.M.; Rao, Y.; Liu, X.; Yin, Z.; Rossi, F. Unraveling the Role of Amino Acid L-Tryptophan Concentration in Enhancing CO_2 Hydrate Kinetics. *Energies* **2024**, *17*, 3702. [CrossRef]
45. Li, Y.; Yin, Z.; Lu, H.; Xu, C.; Liu, X.; Huang, H.; Chen, D.; Linga, P. Evaluation of amino acid L-leucine as a kinetic promoter for CO_2 sequestration as hydrate: A kinetic and morphological study. *J. Environ. Chem. Eng.* **2023**, *11*, 111363. [CrossRef]
46. Kumar, A.; Bhattacharjee, G.; Barmecha, V.; Diwan, S.; Kushwaha, O.S. Influence of kinetic and thermodynamic promoters on post-combustion carbon dioxide capture through gas hydrate crystallization. *J. Environ. Chem. Eng.* **2016**, *4*, 1955–1961. [CrossRef]
47. Hassanpouryouzband, A.; Yang, J.; Tohidi, B.; Chuvilin, E.; Istomin, V.; Bukhanov, B. Engineering, Geological CO_2 Capture and Storage with Flue Gas Hydrate Formation in Frozen and Unfrozen Sediments: Method Development, Real Time-Scale Kinetic Characteristics, Efficiency, and Clathrate Structural Transition. *ACS Sustain. Chem. Eng.* **2019**, *7*, 5338–5345. [CrossRef]
48. Linga, P.; Kumar, R.; Englezos, P. Gas hydrate formation from hydrogen/carbon dioxide and nitrogen/carbon dioxide gas mixtures. *Chem. Eng. Sci.* **2007**, *62*, 4268–4276. [CrossRef]
49. Nesterov, A.N.; Reshetnikov, A.M. New combination of thermodynamic and kinetic promoters to enhance carbon dioxide hydrate formation under static conditions. *Chem. Eng. J.* **2019**, *378*, 122165. [CrossRef]
50. Qureshi, M.F.; Khandelwal, H.; Usadi, A.; Barckholtz, T.A.; Mhadeshwar, A.B.; Linga, P. CO_2 hydrate stability in oceanic sediments under brine conditions. *Energy* **2022**, *256*, 124625. [CrossRef]
51. Guerra, A.; McElligott, A.; Yang Du, C.; Marić, M.; Rey, A.D.; Servio, P. Dynamic viscosity of methane and carbon dioxide hydrate systems from pure water at high-pressure driving forces. *Chem. Eng. Sci.* **2022**, *252*, 117282. [CrossRef]

52. Mohammadi, A.; Pakzad, M.; Mohammadi, A.H.; Jahangiri, A. Kinetics of (TBAF + CO_2) semi-clathrate hydrate formation in the presence and absence of SDS. *Pet. Sci.* **2018**, *15*, 375–384. [CrossRef]
53. Lv, X.; Zhang, J.; Zhao, Y.; Liu, Y.; Xu, J.; Ma, Q.; Song, S.; Zhou, S. Experimental Study of the Growth Kinetics of Natural Gas Hydrates in an Oil–Water Emulsion System. *ACS Omega* **2022**, *7*, 599–616. [CrossRef] [PubMed]
54. Li, Y.; Wu, P.; Sun, X.; Liu, W.; Song, Y.; Zhao, J. Creep Behaviors of Methane Hydrate-Bearing Frozen Sediments. *Energies* **2019**, *12*, 251. [CrossRef]
55. Zhang, Z.; Li, Y.; Zhang, W.; Wang, J.; Soltanian, M.R.; Olabi, A.G. Effectiveness of amino acid salt solutions in capturing CO_2: A review. *Renew. Sustain. Energy Rev.* **2018**, *98*, 179–188. [CrossRef]
56. Javidani, A.M.; Abedi-Farizhendi, S.; Mohammadi, A.; Hassan, H.; Mohammadi, A.H.; Manteghian, M. The effects of graphene oxide nanosheets and Al_2O_3 nanoparticles on the kinetics of methane + THF hydrate formation at moderate conditions. *J. Mol. Liq.* **2020**, *316*, 113872. [CrossRef]
57. Ahmadi, H. Measurement and Modeling of Carbon Dioxide Solubility in Polar and Nonpolar Solvent. *Res. J. Appl. Sci. Eng. Technol.* **2012**, *4*, 2357–2360.

processes

MDPI

Review

Gas Hydrate Distribution Influenced by Sea Floor Slide and Canyon Erosion in the Shenhu Slope, Northern South China Sea Margin

Yuhang Huang, Xinghe Yu * and Chao Fu

School of Energy Resources, China University of Geosciences, Beijing 100083, China; dr_yuhanghuang@163.com (Y.H.); fuchao@cugb.edu.cn (C.F.)

* Correspondence: billyu@cugb.edu.cn

Abstract: The Shenhu area, located on the northern continental slope of the South China Sea, is a confirmed gas hydrate-enriching region, but the sedimentary unit, causative mechanisms, and evolution processes of the strata that contain hydrate remain unclear. Furthermore, the recognition of bottom simulating reflectors (BSRs) rests on qualitative description; there is no quantitative method for the characterization or three-dimensional depiction of BSRs. This review examines the sedimentary features and key factors controlling gas hydrate distribution in the region, based on high-resolution sequence stratigraphy combined with drilling and logging data from hydrate drilling projects in the Shenhu area. The main findings of this study include (1) BSRs are mainly distributed in the ridges of the continental slope and in the slip blocks of the side slope, with hydrates developing along a thin layer (10–40 m) below the hydrate stability zone, as confirmed by drilling results; (2) The distribution of BSRs is strongly influenced by the presence of gas chimneys, the migration of deepwater channels, and the erosion and sedimentation processes of canyons, all of which are directly or indirectly related to the accumulation, distribution, and formation of hydrate reservoirs; (3) The sand factor is generally less than 10%, and BSRs accumulate in areas where the sand factor is higher (4–10%). Hydrate saturation shows a positive correlation with porosity. This research also identifies the early Pleistocene erosion/resedimentation event as a key factor that controls the non-homogeneous distribution of hydrates in the region. This research highlights the role of deepwater canyon erosion and slumping processes in controlling gas hydrate formation, offering new insights into the impact of dynamic geological processes on hydrate accumulation. This study provides valuable knowledge for future hydrate exploration and global resource assessments.

Keywords: the northern continental slope of the South China Sea; the Shenhu slope; gas hydrate; bottom simulating reflectors (BSRs); sedimentation feature; canyon; submarine landslide

1. Introduction

Gas hydrate, which is also called flammable ice, is a newly discovered clean energy with great potential [1–3]. It is a near-solid substance formed by water and natural gas under low-temperature and high-pressure conditions. In deepwater areas, it occurs at depths deeper than 300 m. In the past 20 years or so, through the international ocean drilling program (ODP) and deep-sea drilling project (DSDP), the occurrence of gas hydrate has been located worldwide [2,4]. The northern continental slope of the South China Sea is a key area for the exploration of gas hydrate in China's ocean waters. In the past 20 years, substantive geologic, geophysical, and geochemical investigations have been conducted in this region, and several gas hydrate prospective areas have been identified [2]. The China Geological Survey successfully obtained samples of gas hydrate in the northern Shenhu slope (GMGS-1) and the Dongsha area (GMGS-2) of the South China Sea in 2007 and 2013, respectively. These samples proved that the northern areas contain hydrate reservoir-formative conditions and exploration prospects. The Shenhu area is currently a key target for the exploratory development of gas hydrate in China [5,6]. The heterogeneous characteristics and extensive spatial distribution patterns of potential hydrate development sites on the Shenhu slope are primarily attributed to sedimentation, particularly the direct relationship with submarine landslides and canyon erosion [7,8].

Studies of gas hydrate have focused on the effects of temperature, pressure, and air source on the formation of hydrate and on the influence of sedimentation on its distribution [9,10]. In particular, the relationships among sedimentary evolution and the formation of hydrate and hydrate reservoirs are "hot topics" in the field. However, despite extensive attention from scholars, many questions remain [11]. For example, although some scholars have summarized the favorable sedimentation conditions for the occurrence of gas hydrate and have simulated the hydrate-formative process [12], previous work has not been confirmed by borehole information. The aging of hydrate, determination of key interfaces, identification of sedimentation units, and causative mechanisms and sedimentary evolution processes of strata that contain hydrate remain sparse in the literature. There is no single accepted methodology for identifying the type of deepwater sedimentary system during the hydrate reservoir formation process, nor is there a shared vocabulary for describing and characterizing the time-spatial distribution of the internal structure of sedimentary bodies or the spatial allocation relationship between different sedimentary bodies and hydrate [13,14]. The drilling and acquisition of hydrate currently depend largely on the identification of bottom simulating reflectors (BSRs). At present, the identification of a BSR still rests on a qualitative description and on tracking; there is no quantitative characterization of the geophysical parameters of strata that contain BSRs. Although previous studies have also proposed biochronologic demarcations of the key geologic interface inside the hydrate drilling area in the Shenhu area, the ages of the strata containing hydrate are disputed [15–17]. Therefore, it is of great importance to understand the temporal and spatial evolution of the sedimentary system under the frame of the sequence stratigraphy to further define the sedimentation features of gas hydrate in this area.

This paper uses the research findings of previous papers, which include published distribution characteristics of BSRs in the gas hydrate drilling area of the Shenhu area and high-resolution (HR) sequence stratigraphic units acquired through drilling and 3D seismic data, to study the effects of sedimentary facies, sand factor, and lithologic composition on the development of BSRs. It also addresses the relationships among the submarine landslide and canyon in the area and the formation and conservation of hydrate, explores the sedimentation features of the hydrate drilling area, and establishes a sedimentation model in this area.

2. Geological Setting

The Shenhu hydrate investigation area is near the southeastern end of the Shenhu submerged shoal, in the middle part of the northern South China Sea margin between the Xisha Ocean Trough and the Dongsha Islands (Figure 1A,B). It lies on the southern wing of the Zhu-II Depression within the Pearl River Mouth Basin, located on the slope of the Baiyun Sag in the PRMB (Figure 1C). The submarine relief inclines from the north to the south as a whole, and the water depth gradually changes from 1000 m to 1700 m. Two nearly north–south-oriented submarine troughs are developed in the northern part of the study area, extending from west to east [18,19]. The western trough, located between 115°06′ and 115°08′, is narrow and elongated with a greater depth. In contrast, the eastern trough, situated between 115°09′ and 115°11′, is wider and less deep compared to the western trough. It can be divided into south and north parts by the underwater contour at 1350 m water depth. The northern part with a water depth of 1000 to 1350 m is steep, with the underwater gradient being generally approximately 3.0×10^{-2}. The southern part, with a water depth of 1350 to 1500 m, is relatively smooth, with the underwater gradient being generally approximately 2% [20,21]. The isobaths in this area are relatively sparse compared to the northern part of the investigation area, trending nearly east–west and gradually transitioning into the abyssal plain. Eight sites were drilled, and hydrate cores were obtained at three sites (Sites a, b, and c). The sites where hydrate was drilled out were mostly in places with depths of 1200–1350 m (Figure 1D). The main landforms developed in this area included sea knoll, submarine valley, and submarine trench [16,22].

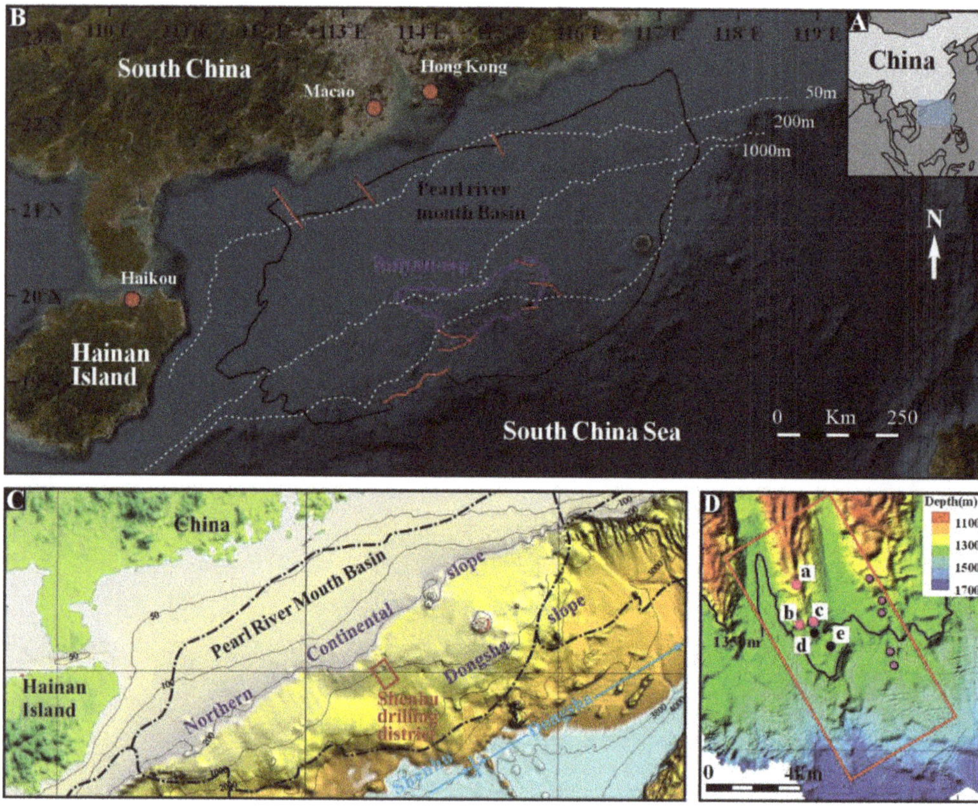

Figure 1. (**A**) Location of the study area; (**B**) Location of Baiyun sag; (**C**) Location map showing the area of gas hydrate exploration in the Pearl River Mouth Basin, the northern South China Sea; (**D**) Bathymetric map of Shenhu drilling district; red dots show cored samples with gas hydrate (Sites a, b, and c), purple and black (Sites d and e) dots show cored samples with no gas hydrate. The water depth ranges from about 800 to 1700 m.

During the Late Oligocene to Early Miocene, the study area was primarily characterized by transitional and marine clastic sedimentation, with the localized development of carbonate platforms and biogenic deposits, which were subsequently overlain by mudstone. After the Middle Miocene, in the northern margin of the South China Sea, the water depth gradually increased from north to south, transitioning from littoral to shallow marine, semi-deep marine, and deep marine clastic sedimentation [21,23] (Figure 2A). Since the Late Miocene, the research area has been in the accelerated sedimentation stage of the structure evolution of the Pearl River Mouth Basin, mainly appearing as a few fracture activities and abnormally high tectonic subsidence velocity [24]. The interfaces of the three secular cycles were T_3, T_2, and T_1 from bottom to top, and the corresponding times were 11.5 Ma, 5.3 Ma, and 1.8 Ma [2,16,22]. The hydrate mainly dispersed over strata sequences IV and III between T_2 and T_1, and the hydrate was drilled out from strata sequence IV at Sites a and b (Figures 2B and 3).

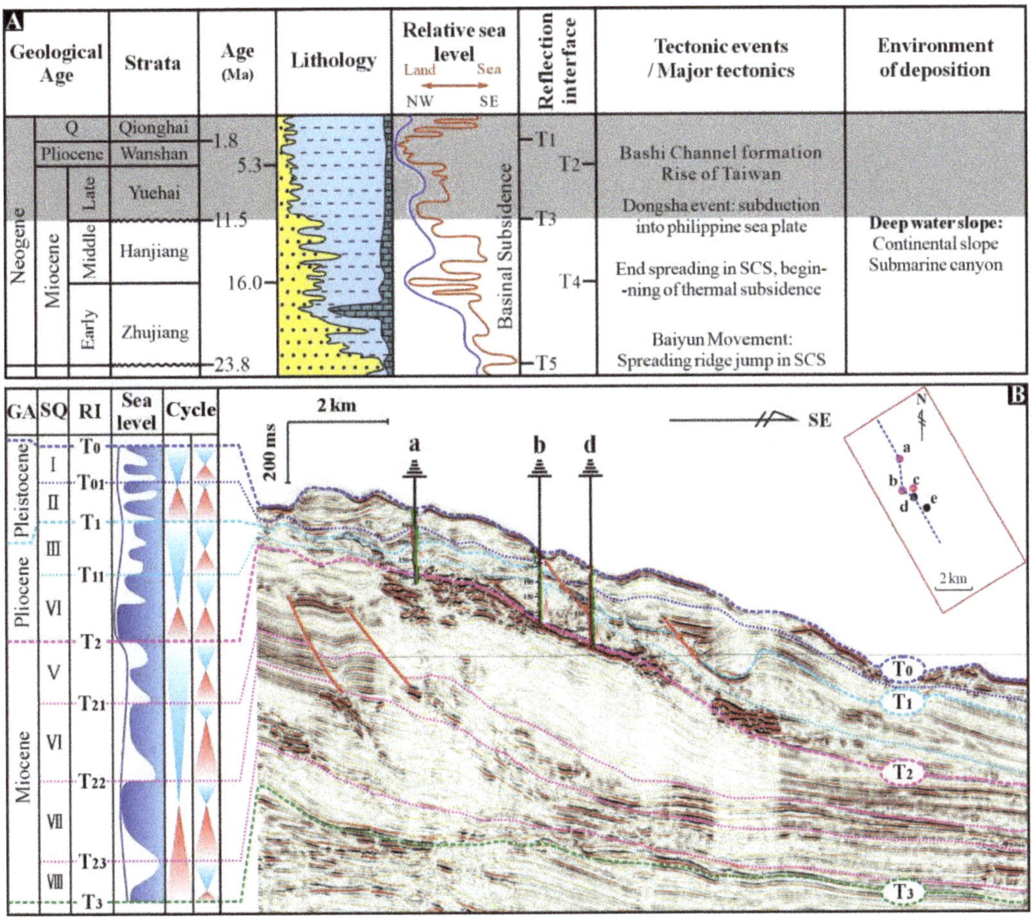

Figure 2. (A) Generalized tectono-stratigraphic column of the Pearl River Mouth Basin; (B) Division of biostratum frame in the drilling strata and sequence stratum of the seismic profile in the hydrate drilling area of the Shenhu slope of the South China Sea.

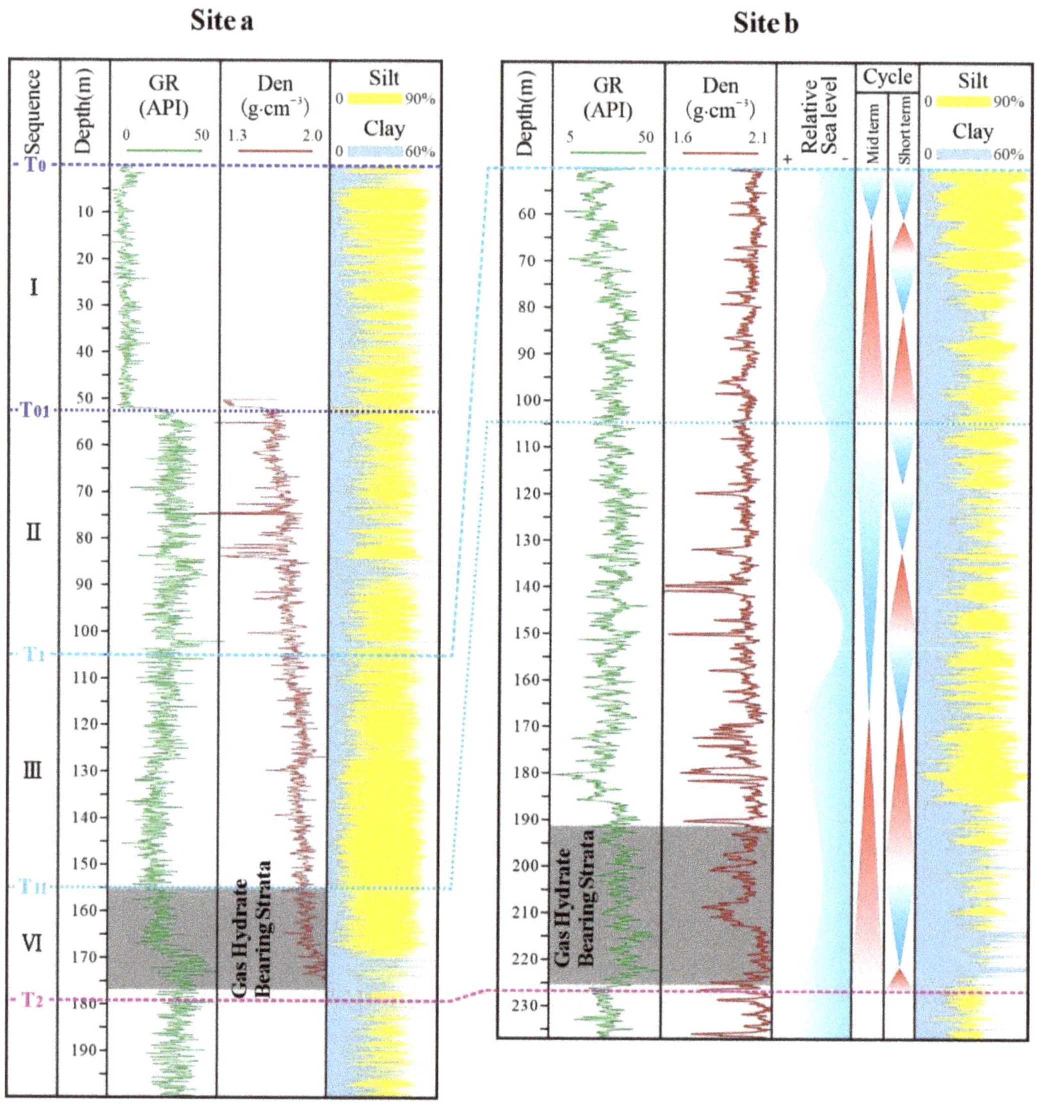

Figure 3. Division of biostratum frame in the drilling strata of Site a and Site b.

3. Data and Methodology

3.1. Seismic and Logging-While-Drilling Data

The 3D seismic data utilized in this study were acquired by the GMGS between 2008 and 2009, employing a frequency of 65 Hz and a sampling interval of 1 ms across an area measuring approximately 12.5 km by 25 km. The logging data were acquired by Fugro and processed by Schlumberger. The data obtained by GMGS included natural gamma radiation, gamma density, resistivity, and sonic velocity. The grain sizes in Sites a, b, and D were measured using a laser particle size analyzer (SYE209).

By leveraging the high vertical resolution of well logging and the high horizontal resolution of seismic data, and, through well seismic calibration in conjunction with the published stratigraphic framework of the study area, an accurate sequence framework can be established [25–27]. Based on the tectonic movement of the South China Sea and the relative change in sea level at the Pearl River Mouth Basin since the Late Miocene, the authors of this paper determined HR sequence divisions in this area using theoretical seismic stratigraphy based on the drilling and logging results from the drilling area in the Shenhu area based on the convergence, divergence, and termination mode of the seismic reflection axis [28,29]. The identification of sequence boundaries was primarily based on seismic characteristics, including reflection frequency, amplitude, continuity, and reflection terminations. Three secular cycles and eight middle-term cycles were identified in the strata since the Late Miocene [15,17]. Based on these, the authors defined eight sequence strata. The seismic boundary is delineated through seismic velocity–lithology analysis and the termination characteristics of reflection waves in the seismic profile. By integrating the sedimentary background of the study area, the conversion relationship between seismic facies and sedimentary facies is established [30].

The processing of drilling data primarily includes sediment grain-size composition analysis, probability cumulative curve analysis, and porosity calculation for cored intervals. This study adopts a modeling approach that relies on the direct conversion of seis-mic data into geological parameters, primarily concentrating on deterministic modeling while incorporating stochastic modeling to develop the geological model of the study area. The root mean square (RMS) amplitude exhibits a positive correlation with rock density [31]. The spatial variation of the RMS amplitude is employed to predict the lateral distribution of lithology, while the extracted RMS amplitude seismic attributes are utilized to estimate sand content. The porosity of the gas hydrate layers at Sites a and b was determined using density logging and subsequently corrected for clay content.

3.2. Sequence Division and Identification of BSR

Previous studies have shown that the formation and occurrence of gas hydrate, as well as the thickness of the gas hydrate stability zone (GHSZ), are affected by sea-level changes [32]. Once gas hydrate mineralizes, it shows distinct features, often seen as geophysical responses and geochemical anomalies, such as bottom simulating reflectors (BSRs) [33]. BSRs are seismic reflections that mark the boundary between hydrate-bearing sediments and free gas-containing sediments. The reflection results from a strong amplitude contrast between high-impedance hydrate sediments and underlying low-impedance sediments. BSRs are the key geophysical indicators of gas hydrate accumulation and mineralization [30,34]. Seismic attributes include geometrical morphology, kinematic characteristics, dynamic features, and statistical characteristics of seismic waves [35]. BSRs have primarily been identified through various anomalies, including abnormally high seismic velocities, dominant amplitudes, blank reflections in seismic profiles, their parallelism with the seabed, and polarity reversals [36,37] (Figure 4). RMS amplitude is the most commonly used seismic attribute in seismic interpretation; it primarily reflects changes in seismic wave energy, emphasizing the strong reflection of a BSR surface and the significant amplitude of hydrate development. Based on the high density of RMS amplitude extracted vertically in the research area, a value-assigned algorithm was utilized to construct a model of seismic attributes [38] (Figure 5).

According to geophysical and logging indicators, the BSRs in the study area are mainly developed in sequences III and IV, between T_2 and T_1, at depths of 150–300 m (Figure 4). On the connected-well profile, the model indicated that reflections below a BSR exhibit strong amplitudes, while those above show weak amplitudes and blank reflections (Figure 5). Strong amplitudes beneath BSRs represent the most significant seismic response characteristic of hydrate in the study area. Above the BSRs, weak amplitude or blank amplitude bands are observed, and most BSRs appear oblique to the geological formations.

These features enable reliable identification of BSRs and facilitate the determination of gas hydrate distribution.

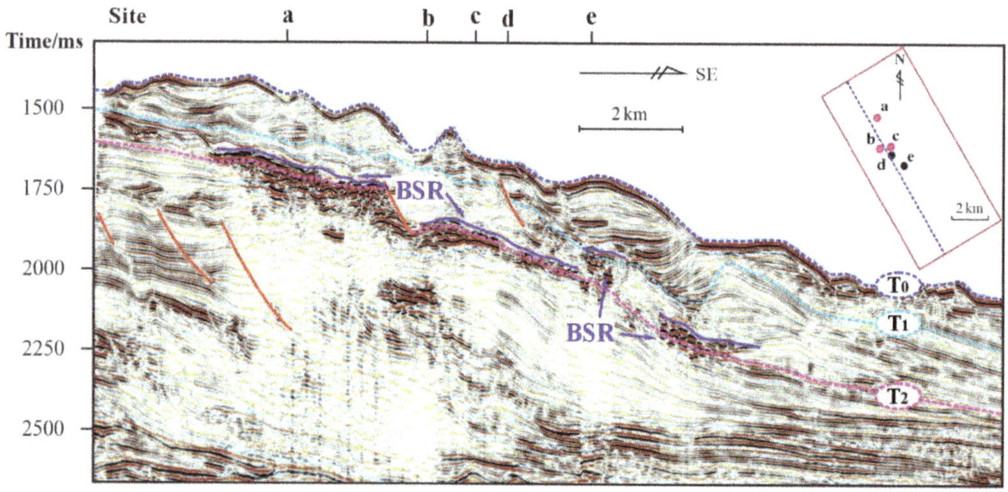

Figure 4. Typical seismic profiles across Sites a–e in the drilling area of the northern Shenhu slope of the South China Sea (the profile position is shown in Figure 1).

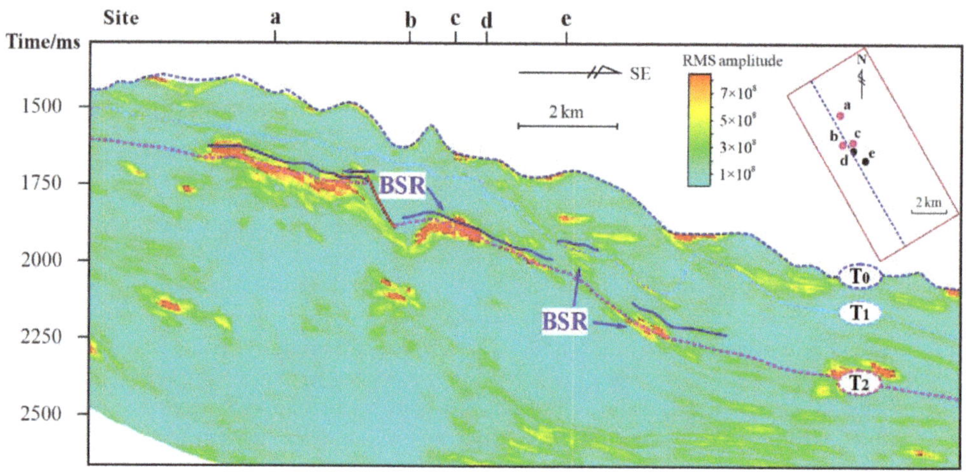

Figure 5. Amplitude attribute of BSRs' typical seismic profiles across Sites a–e in the drilling area of the northern Shenhu slope of the South China Sea.

Judging from the BSR distribution depicted in strata sequence IV and strata sequence III (Figures 1 and 6), BSRs mainly distribute along the ridge of the continental slope and the side slope; most BSRs are located at a water depth of 1200 to 1350 m. Strata sequence IV has four BSR distribution areas, mainly in the northwest and northeast of the research area. Their sizes vary noticeably. According to the drilling results, the strata containing hydrate all lie in this strata sequence. Strata sequence III only has one BSR distribution area, which is large and has good continuity; it mainly lies in the southeast of the research area, distributing in the form of ribbons. Drilling results indicate that hydrate is uniformly distributed within fine-grained sediments on the Shenhu slope, exhibiting vertical stratification and planar zoning.

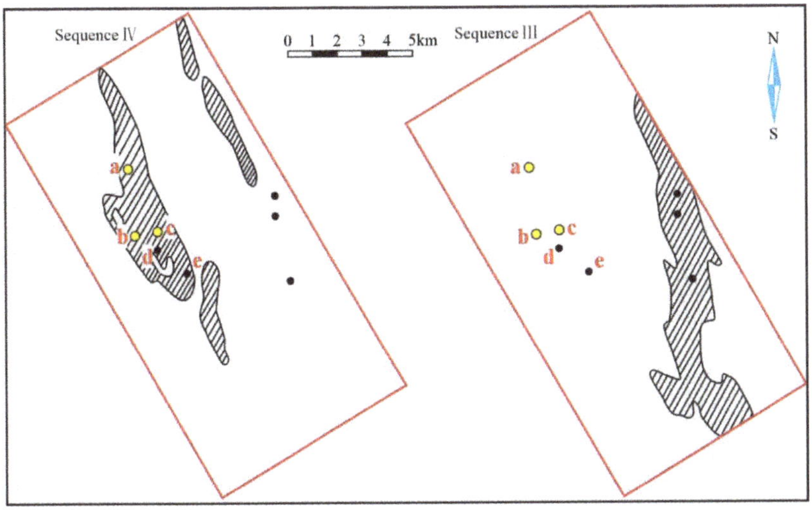

Figure 6. Planar distribution map of BSRs in sequences IV and III in the Shenhu slope.

4. Results

4.1. Seismic Facies and Depositional System Analysis

Seismic facies within the Shenhu slope are classified into six major types (Facies 1 to Facies 6), based on reflections, amplitude, continuity, and external geometry (Figure 7). The study area is characterized by a typical slope gravity flow sedimentation setting. Based on the distinct characteristics of seismic facies and the sedimentary environment, seismic facies are interpreted as corresponding sedimentary facies [39].

Facies 1 is a lens chaotic facies characterized by horizon migration reflections, moderate to high amplitude, moderate continuity, and lens-like external geometry. It exhibits distinct downcutting features, with well-defined boundaries on both sides and the bottom. The interior consists of low-continuity chaotic reflections, suggesting that the filling is composed of coarse-grained or disordered accumulation, with noticeable local side-accretion filling (Figure 7). This suggests that it represents a sedimentary body formed by watercourse filling through multiple sedimentation processes. These channels (CH) are typically developed within canyons and serve as key sediment transport pathways and depositional areas [40,41].

Facies 2 is an irregular chaotic facies characterized by random reflections, high amplitude, low to moderate continuity, and irregular-like external geometry. It is located on the side of the waterway, lacking a defined shape or clear boundary, with a greater thickness and limited lateral extension (Figure 7). The internal reflection features are primarily chaotic and are mainly distributed along the bottom and sides of the canyon, influenced by the canyon's concave shape, resulting in relatively stable morphology. Mass transport deposits (MTDs) represent a small-scale slide occurring within the interior and ridges of the canyon, reflecting an MTD event during the sedimentary period [42,43].

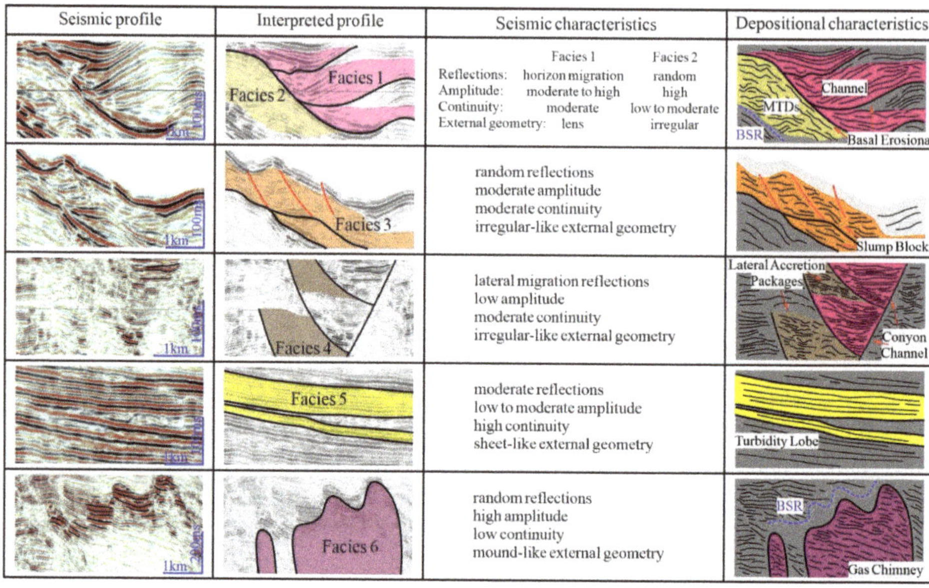

Figure 7. Characteristics of seismic facies with respect to contact relationship with seismic events. Six major facies identified within the study area, based on seismic reflections, amplitude, continuity, and external geometry of seismic reflections.

Facies 3 is an irregular chaotic facies characterized by random reflections, moderate amplitude, moderate continuity, and irregular-like external geometry. It is primarily distributed along the ridge of the continental slope, with a limited distribution range, separated by multiple faults, and exhibiting a typical step-like shape. While its characteristics remain continuous, the influence of faults leads to the formation of multiple sliders, whose boundaries display significant deformation (Figure 7). This facies is interpreted as a mixed system of coarse clastic and argillaceous sediments, with localized slipping and dislocation on the slope, indicative of a slump block (SB) [44,45].

Facies 4 is an irregular chaotic facies characterized by lateral migration reflections, low amplitude, moderate continuity, and irregular-like external geometry. On the side of the channel, several strata sets are inclined toward the axis of the canyon, migrating unidirectionally along with it (Figure 7). This results from lateral accretion packages (LAPs) formed by isobaric current activity in the context of lateral canyon migration in the study area [46,47].

Facies 5 is a sheet subparallel facies characterized by moderate reflections, low to moderate amplitude, high continuity, and sheet-like external geometry. The internal reflection is intermittent, primarily distributed in the flat area at the bottom of the slope, with considerable lateral extension and minimal thickness variations (Figure 7). This suggests a more uniform internal material composition, characteristic of fan- or leaf-shaped lobes (LB) [28,44].

Facies 6 is a mound chaotic facies characterized by random reflections, high amplitude, low continuity, and mound-like external geometry. It is a chaotic internal reflector package, characterized by a chimney-like reflection boundary, appearing as a near-vertical column that gradually expands upward, penetrating multiple sedimentary layers. The bottom exhibits a lack of reflection and blurring, while the top shows a high-amplitude reflection (Figure 7). In the gas migration pathway, disturbance of the surrounding sedimentary layers causes their rearrangement and fracturing. These are gas chimneys (GC) formed as gas migrates upward from deep rock formations [48,49].

Studies have shown that the sea floor gravity flow sediment (especially the contour current sediment and turbidity current sediment), the slump fan of the delta front, and the slope fan at the turn of the structure are the favorable facies belts for the development and occurrence of gas hydrate [2]. BSRs are distributed in strata sequence IV and strata sequence III. These strata sequences mainly developed a channel, natural barrier, lobe, moderate sea sediment, and slip block (Figure 8). Parts of the strata were eroded by the deuteric deepwater channel. The channel was approximately 1 km wide in the northwest. As the terrain smoothed to the southeast, the channel branched; its tortuosity factor increased, and it became much narrower. At its end, the channel developed a lobe. The slip block mainly developed on the sea knoll side slope; thus, this area formed the obvious slumping-eroded channel-lobe sedimentary system [50].

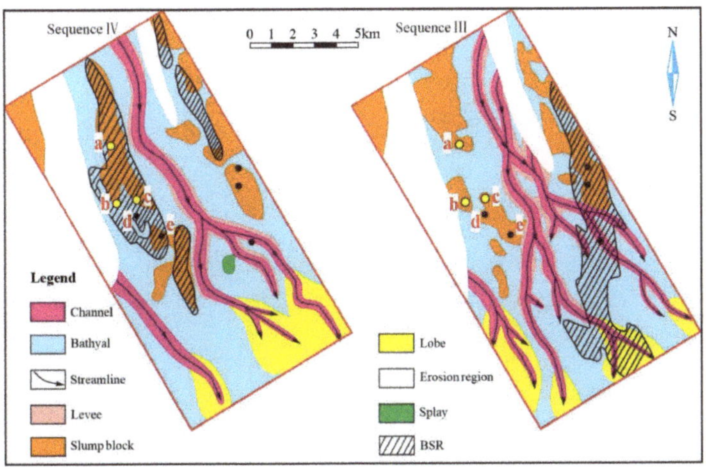

Figure 8. Distribution diagram of sedimentary facies and BSR in strata sequence IV, revised from the literature; distribution diagram of sedimentary facies and BSR in strata sequence III, revised from the literature [39].

4.2. Sand Content Model

From a statistical perspective, RMS amplitude correlates positively with rock density [31]. Its variance in a plane will reflect the lithologic change. Therefore, the statistical relationship between the seismic attribute and lithologic characteristics of drilling could be established according to the plane variance and analysis of RMS amplitude as well as the drilling data. The correlation between the seismic attribute and lithologic characters of drilling can be obtained based on the well bypath RMS amplitude and sand factor of the corresponding coring section above the single well, and then the sand factor can be calculated according to the RMS amplitude seismic attribute to draw the sand factor plane view. The sand factor in the research area is low as a whole, essentially below 10%. BSRs in strata sequences IV and III are mainly distributed in the relatively high-value section. The sand factor in the three sites (a, b, and c) where hydrate was drilled in strata sequence IV is relatively high (4–10%). The nearby Site d is in the low-value section, from which no hydrate was drilled (Figure 9).

The sand content of sequence IV in the study area generally exceeds 1.0%, with areas where the sand content exceeds 2.7%, forming northeast–southwest-oriented bands, which show a strong correlation with the sedimentary facies of sequence IV described in

Section 4.1. A comparison with the sedimentary facies map reveals that the sand content in the lobes and channels ranges from 2.7% to 4.0%, with some areas exceeding 4.0%, and the sand content in the slump blocks is generally greater than 4.0% (Figure 10). The sand content of sequence III generally exceeds 1.2%, with areas where the sand content exceeds 2.7% forming east–southwest-oriented bands, which also show a strong correlation with the sedimentary facies of Sequence III described in Section 4.1. A comparison with the sedimentary facies map shows that the sand content in the lobes and channels ranges from 2.7% to 7.6%, while the sand content in the slump blocks generally exceeds 5.0%, with some areas exceeding 7.6% (Figure 10). The sand factor, which is closely related to the lithology and texture of the sediments, is a critical parameter in understanding the spatial distribution of gas hydrate. The high RMS amplitude values correspond to areas with greater density contrasts between the sand and surrounding sediments, helping to identify key regions where hydrate accumulation is more likely [51,52].

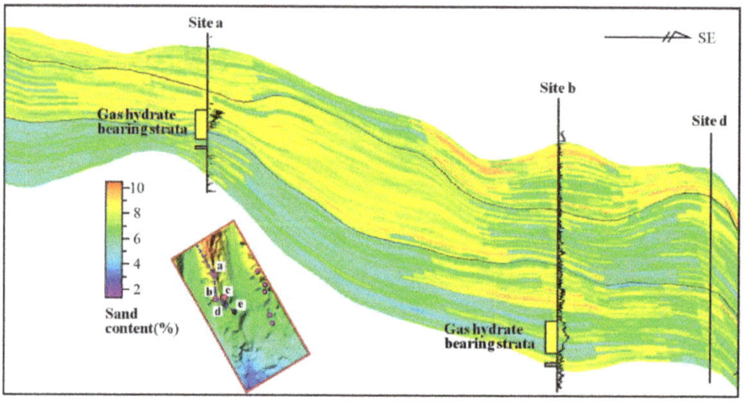

Figure 9. Section plane of cross-hole sand content.

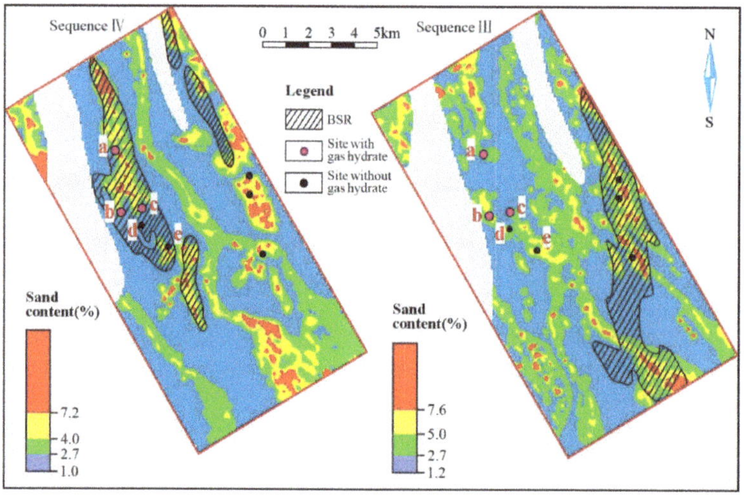

Figure 10. Sand content and BSR distribution in strata sequence IV. Sand content and BSR distribution in strata sequence III.

4.3. *Relevant Parameters of the Strata That Contain Hydrate*

4.3.1. Lithologic Characteristics

The hydrate samples drilled in the research area are distributed evenly in the pores of fine-grained clay silty sand that is rich in foraminifera and in calcic nannofossils in the dispersal or conglutination form. A component analysis of the size grade of core intervals in this study suggests that the sediment type is mainly composed of clayey silt and silty sand. The hydrate mainly distributes in fine-graded sediments in lentoid, concretion, grain, and flaky forms. Their granularity is fine, but they distribute widely between 1Φ and 11Φ. The sand content varies from 1.0% to 10% (Figure 11). Although the rough component is a small percent of the whole, it can increase the porosity and penetrability of sediments, which is good for the accumulation and occurrence of gas hydrate [2,16].

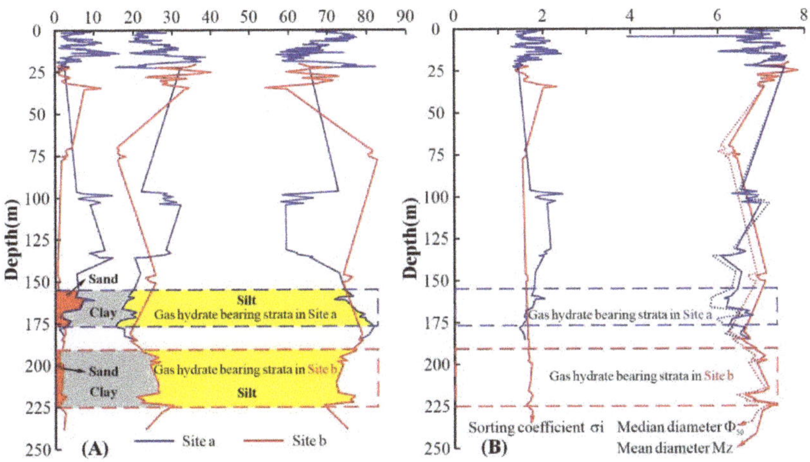

Figure 11. Granularity characteristic curve of the site that contains hydrate (the part in the box is the hydrate-bearing stratum). (**A**). Grain size composition content (%) of the core sections from Sites a and b; (**B**). Grain size distribution curves for Sites a and b, including sorting coefficient, mean diameter, and median diameter.

The probability cumulative curve is the curve that is most widely used in sedimentology to analyze the hydrodynamic conditions for deposit formation and further distinguish the depositional environment through its characteristics [53]. The particle size probability cumulative curves of the debris channel at Sites a and d exhibit a two-stage jumping population with a high-slope–low-slope pattern and a high-slope suspension population, with the cutoff point of the S curve occurring between approximately 3 and 4. The slump body demonstrates a typical single-stage, high-slope suspension population. The natural barrier appears as low-slope saltation population and high-slope suspension population, and the S truncation point is at approximately 5~6; the bathyal mud appears as "upper arc-like" one-section-type suspension population. All these granularity probability curves are mostly suspension population with a lower slope, which reflects the typical gravity flow characteristic [54,55] (Figure 12).

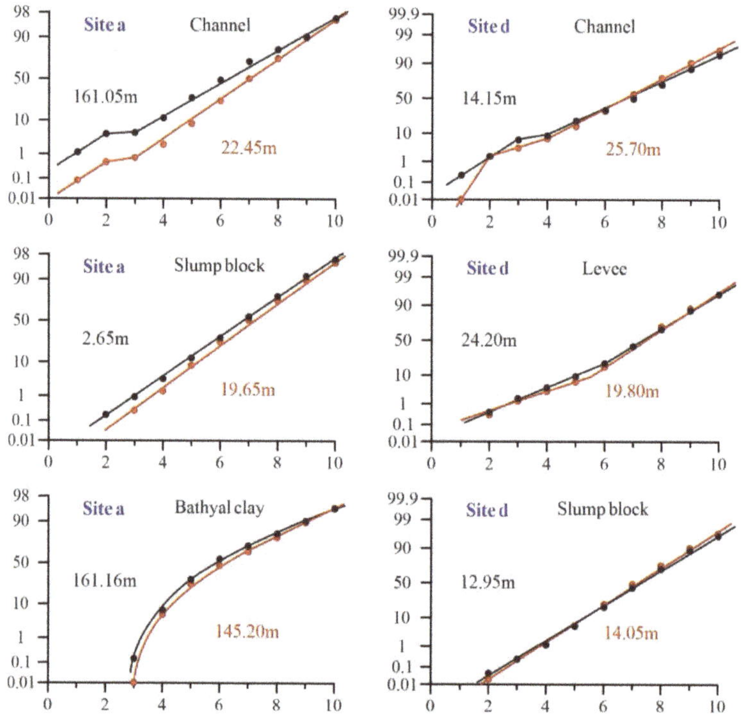

Figure 12. Granularity probability curve at typical sites in the research area.

4.3.2. Reservoir Porosity

A point projection of the porosity (calculated through density logging, after shaliness correction) of the hydrate-bearing strata and hydrate saturation (calculated through the resistivity and Archie equation) at Sites a and b is made (Figure 13).

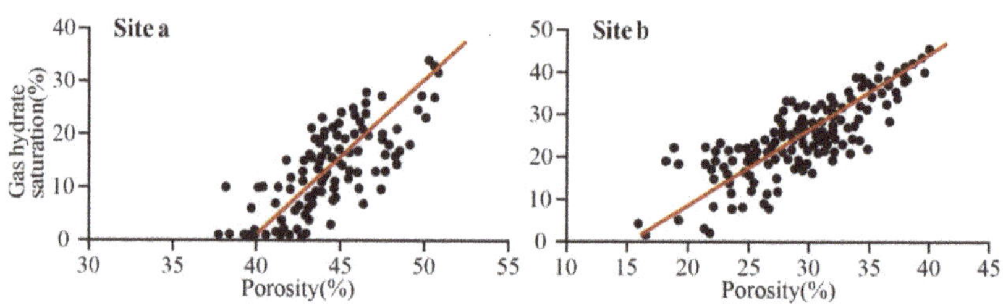

Figure 13. Convergence of porosity and saturation of the hydrate strata at Sites a and b.

When the sediment porosity is higher and the capillary pressure is lower, the methane fluid can more easily migrate to the reservoir bed for enrichment and mineralization. It is observed that higher porosity and higher hydrate saturation are better for hydrate formation [56]. Studies detailing the impacts of the physical properties of the reservoir bed on the formation and distribution of hydrate according to hydrate cores and logging data were acquired, thus helping to forecast the advantageous distribution areas of hydrate. Many hydrates found in the world form relatively coarse sediments; coarse sediments usually have greater porosity. For example, hydrate forms in the turbidite sandstone sandwich layer in the bathyal facies' mudstone at Site U1326 in north Cascadia, and there is no hydrate in the mudstone [57].

5. Discussion

5.1. Factors Influencing the Distribution of Gas Hydrate and BSRs

In addition to the major temperature and pressure conditions, the formation and occurrence of gas hydrate also require specific sedimentation conditions. For example, the sediment must have good porosity and relatively poor penetrability, so as to maintain the water, reservoir space, and preservation environment to form the gas hydrate; the reservoir bed must possess a certain thickness; and there must be adequate air supply and a relatively high rate of sedimentation [2,58,59]. Furthermore, fractures, diapiric folds, mass movement, subduction zones, and tectonic subsidence velocity are of great importance to the existence of gas hydrate [24,60,61]. In order to meet these requisite conditions, the following additional conditions must be met by sedimentation: the sand/mud content in the sedimentation must be appropriate, the rate of sedimentation must be relatively high, and the sedimentary bodies must be thick. When these conditions are met, the fan bodies, deltas, and gravity flow sedimentary systems are considered favorable. The study of sedimentary system characteristics like distribution, sand factor, and lithologic composition is helpful for analyzing the sedimentation conditions and controlling factors of hydrate reservoir formation [7].

5.1.1. Relationship Between Sedimentary Facies and BSRs

The superposition of the sedimentary facies and BSR distribution range showed that BSR distribution had a close relationship with the slip block in strata sequence IV. Under the sea, an earthquake, volcanic eruption, storm surge, fast accumulation of sediment, or excessive inclination of slopes can cause slump. Loose and richly organic slip-block sediments can cause an excessive release of fluid, owing to the lateral compacting action. Gases can spread to the superficial parts and form hydrate through diffusion. In Sequence III, the majority of BSRs are concentrated in the slump blocks, with smaller portions distributed in the bathyal, lobes, and channels. This indicates that the distribution of BSRs is closely related to the slump blocks within the depositional facies, although the slump blocks are not the only controlling factor (Figure 8).

5.1.2. Sand Content in Hydrate-Bearing Strata

The distribution of BSRs in the study area overlaps significantly with areas of high sand content. Sequence III's BSRs are primarily distributed in areas where sand content exceeds 4.0%, while Sequence IV's BSRs are primarily located in areas where sand content exceeds 5.0% (Figure 10). The sand factor in the three sites (a, b, and c) where hydrate was drilled in strata sequence IV is relatively high (4–10%). The nearby Site d is in the low-value section, from which no hydrate was drilled, showing that the formation of hydrate needs certain reservoir space [62] (Figure 9).

Sand content effectively reflects the sedimentary pattern and directly influences the development of reservoir space and porosity, which in turn affect the formation of gas hydrates [63,64]. The model simulation of sand content aligns well with the distribution of well data. Higher sand content in the hydrate interval leads to higher hydrate saturation. Gas hydrate primarily forms and accumulates in the center of sand layers, with minimal or no accumulation at their top, bottom, or edges. Gas hydrate reservoirs require adequate porosity, typically provided by high sand content. As gas ascends via gas chimneys and reaches high-sand-content horizons, it migrates horizontally and is laterally injected into thick hydrate stability zones. The surrounding low-sand-content mudstone provides an effective cap [65].

However, several other locations also fall within the high-value segment, yet no gas hydrate was found (Figure 9), suggesting that the formation of hydrate is influenced by factors beyond the availability of reservoir space (e.g., reservoir formation environment, reservoir properties, and gas hydrate source are also required) [66]. The relationship between the RMS amplitude and the sand content is pivotal in mapping these sedimentary environments (Figures 9 and 10). The detailed correlation between sand content and sedimentary facies, especially within slump blocks and the lobe areas, is essential for predicting the distribution and concentration of gas hydrates [67] (Figures 8 and 10).

5.1.3. Petrophysical Properties in Hydrate-Bearing Strata

The silt content in gas hydrate-bearing strata is typically high, and there are no specific requirements for the granulometric distribution characteristics [68]. However, the top of the gas hydrate-bearing strata is characterized by fine-grained sediments. The granularity characteristic curve shows that, with an increasing depth of gas hydrate-bearing strata at Site b, the sorting coefficient remains relatively stable, while both the median and mean diameters gradually increase. In contrast, at Site a, as the depth of gas hydrate formation increases, the sorting coefficient gradually decreases, while the median and mean diameters also increase (Figure 11). The sorting coefficient between the gas hydrate-bearing strata and the overall strata in the study area is relatively high (around 2), suggesting that the sedimentary environment is dynamic. Based on the sedimentary characteristics of the study area, it can be inferred that these strata were formed in a deepwater setting influenced by slump processes [69,70]. Notably, the median and mean diameters of the overlying strata containing the gas hydrate-bearing strata tend to decrease, which indicates that fine-grained sediments are particularly favorable for gas hydrate storage [71]. As the local slump movement weakens, finer-grained sediments are gradually deposited, which is conducive to the enrichment of gas hydrates. However, the accumulation of gas hydrates is localized, and not all strata with this characteristic can form gas hydrate reservoirs. The stable zone for gas hydrate typically occurs at deep water depths, where high-pressure and low-temperature conditions prevail [72]. As a result, gas hydrate accumulation zones can form in environments with sufficient depth, offering favorable storage spaces and capping conditions.

Gas hydrate was encountered in both the 161.05 m channel sand body and the 161.16 m bathyal clay at Site a. The three-stage probability cumulative curve of the 161.05 m channel sand body at Site a consists of three distinct sections: a high-slope fine particle section, a gentler medium particle size section, and a coarse particle section that increases again. At low water flow speeds, fine particles are prone to settling, forming the high-slope fine particle section. At a certain flow rate, medium-sized particles (e.g., sand particles) gradually begin to deposit as the flow rate decreases. As the water flow speed further decreases or the deposition environment changes, larger particles, such as gravel and coarse sand, begin to deposit, forming the rising section of the curve. This indicates the presence of traction, jumping, and suspension loads, which are characteristic of high-energy environments [73,74]. The upper arc of the 161.16 m bathyal clay probability cumulative curve reflects the gradual accumulation of coarser particles during sedimentation, typically concentrated in the upper part of the sediment or confined area. This suggests that the sediment particle size increases as water flow intensity weakens. In general, gas hydrates are predominantly found in strata formed under strong hydrodynamic conditions [75].

The reservoir with larger particle sizes and relatively high sand content typically exhibits a well-developed pore structure and enhanced permeability (Figure 13). Fine particles, such as clay, tend to occupy the pore spaces between larger particles, thereby reducing the overall porosity of the reservoir [76]. A higher sand content effectively mitigates the infilling of these fine particles, thereby preserving a larger pore space. Larger particle sizes are generally correlated with increased permeability, which facilitates the easier flow of fluids, such as gas hydrates, through the reservoir [77].

5.1.4. The Influence of Gas Chimneys on the Distribution of BSRs

Gas hydrate and BSRs are closely related to the distribution of gas chimneys. Gas hydrates occur largely near mud volcanos or the tops of mud diapirs [78]. Diapiric structures appear as arcs, strong or weak cut-throughs, gas chimneys, diapiric folds accompanied by faults or fractures, and sea floor pockmarks. A gas chimney is seismically imaged as the abnormal reflection formed at a seismic profile owing to the vertical migration of natural gas (or fluid) (Figure 14). It forms owing to the excess pressure of the gas pool, structural low stress, and synthetic actions of the sealing layer of mud shale [79]. The gas chimney in the research area mainly developed in the relatively high regions (canyon ridge), which are not eroded by the canyon (Figure 15). Gas chimneys serve as the primary conduits for the upward migration of gas hydrate and are essential for gas reservoir formation [80]. Bottom simulating reflectors (BSRs) are typically located either directly or diagonally above gas stacks (Figure 14, Figures 16–19). Moreover, in the planar view, the overall area occupied by BSRs and gas hydrate is larger than that of the gas stack (Figure 15), suggesting that gas hydrate undergoes lateral migration upon reaching an environment conducive to reservoir formation via the gas chimney.

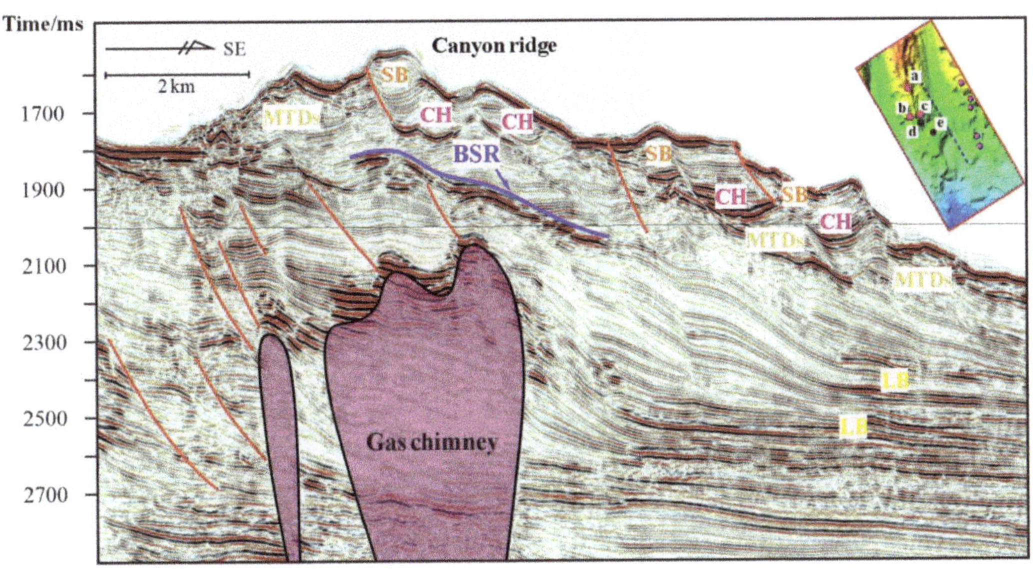

Figure 14. BSRs are prevalently distributed near the top of mud diapirs or gas chimneys. A diapiric fault often observed in and around a diapir. They (gas chimneys and faults) are good channels for the upward migration of gas.

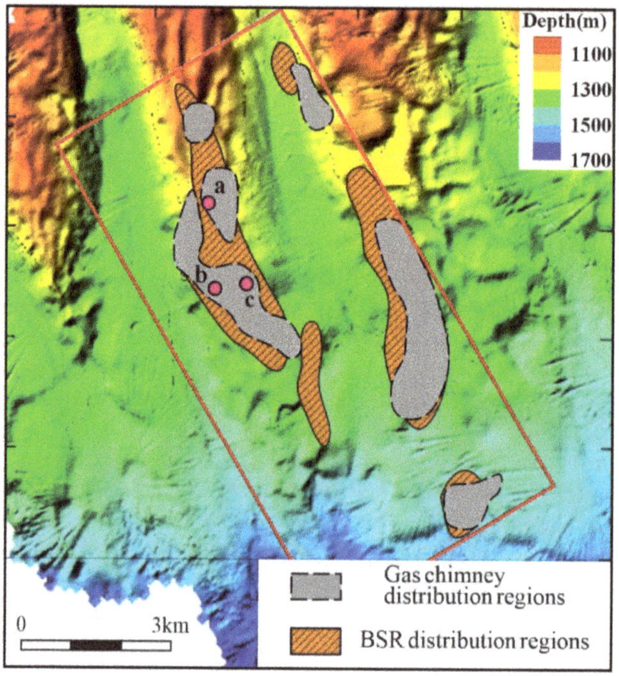

Figure 15. Gas chimney distribution and its overlap with BSRs in the hydrate drilling zone.

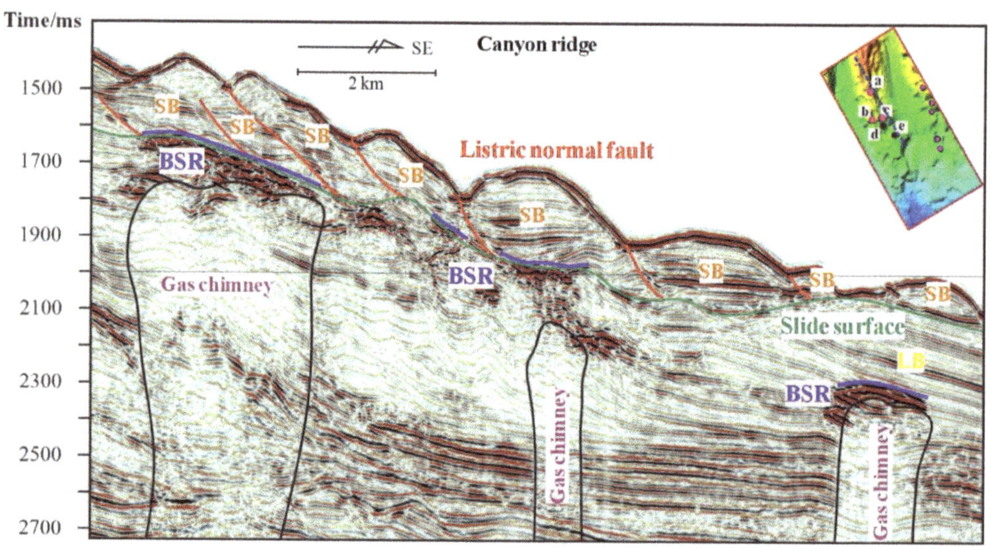

Figure 16. Sliding mass and BSR in the hydrate drilling area of the Shenhu slope.

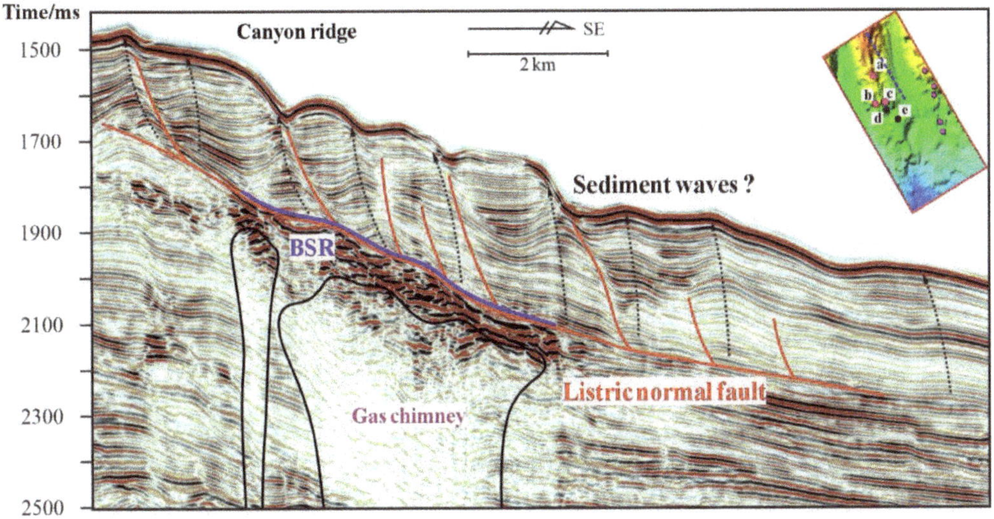

Figure 17. Suspected sediment waves and BSR in the hydrate drilling area of the Shenhu slope.

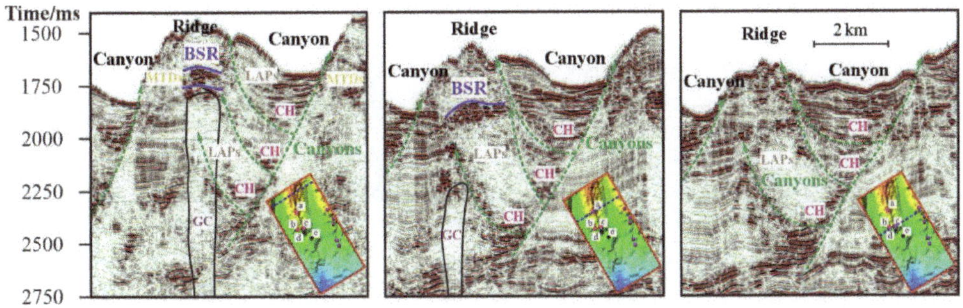

Figure 18. Characteristics of continental slope canyons in the Shenhu slope of the South China Sea.

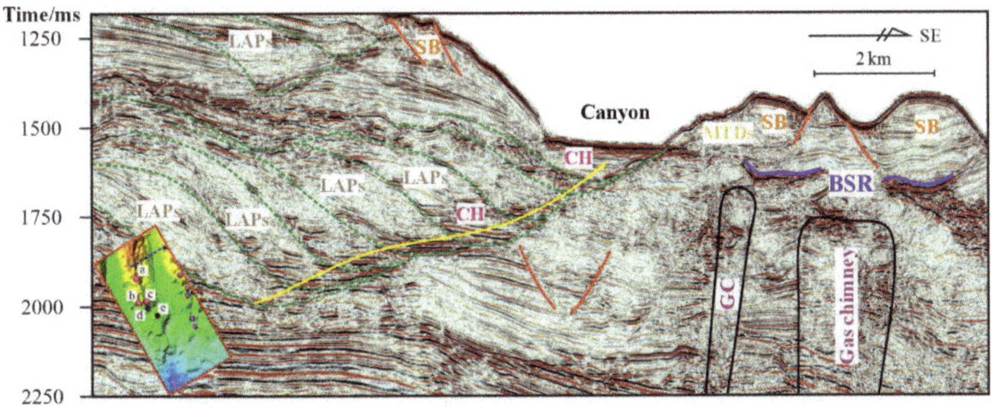

Figure 19. Lateral migration characteristics of continental slope canyons in the Shenhu slope of the South China Sea.

5.2. Origin of Favorable Reservoirs for Gas Hydrates

5.2.1. Submarine Landslide

Submarine landslides are massive movements, under which mass flow and hyperpycnal flow are formed directly or indirectly by paroxysmal factors. The major functional processes include sliding and slumping. The development of slip blocks provides a relatively appropriate temperature and pressure environment for the occurrence of gas hydrate. A slip block might be an induction factor of the slump structure, or slumping might be a consequence produced by the resolution of gas hydrates [81]. Submarine landslides are closely related to the development of hydrate. BSRs are consistently distributed at the base of slump blocks [82].

The upward migration of free gas through gas chimneys has contributed to the destabilization of the preexisting sediments on the slope. This destabilization, driven by episodic events such as seismic activity and ocean currents, enhances the susceptibility of the sediments to sliding [47,83]. The formation of slump blocks provides favorable conditions for gas hydrate accumulation. The repeated slippage of these blocks leads to strata with higher sand content and poorer sorting. As slumping activity diminishes, the sedimentary environment stabilizes, allowing finer-grained deepwater sediments to form more effective cap layers for gas hydrate reservoirs [84,85]. BSRs are primarily located in the lower part of the sliding mass and slip block, adjacent to the sliding surface. The upper strata exhibit a series of plow-type faults that terminate at the sliding surface (Figures 16 and 17). The sliding surface is characterized by sediment liquefaction and sufficient fluid activity on a large and continuous scale, with gas hydrates accumulating in a relatively narrow belt above the BSRs. This is consistent with the results discussed in Section 5.1. Mass flows resulting from slip–slump processes evolve into turbidity currents, depositing multi-stage lobes on the deepwater plain. Compared to the surrounding deep-sea fine-grained sediments, these lobes exhibit higher sand content and greater potential for reservoir formation. Additionally, the well-sorted particle size and low permeability of these deposits often enable them to act as effective cap layers for hydrate reservoirs [86,87]. This facilitates gas accumulation within the hydrate stability zone, thereby promoting the formation of gas hydrates. Thus, as long as there is a sufficient amount of free gas at the base, coupled with gas chimneys, a viable gas hydrate reservoir can form.

Some scholars deduced from characteristics of the seismic profile that there were sediment waves in this area (Figure 17) and believed that the area was composed of a series of sediment waves that upslope migrated, thus forming the reverse sedimentary structure [88]. The waveform is short and steep on the upslope sides, but is long and gentle on the downslope flanks. There is good correspondence between the trough of the sediment wave and the sliding surface in the strata, which suggests the cause of gravity deformation in the wave field. But the author does not think it has the typical characteristic of sediment waves with "sill-like waveform and uniform thickness"; furthermore, its distribution range is very limited, which suggests that it belongs to the sliding mass.

5.2.2. Canyon and Deepwater Channel

Depending on the influence of erosion or deposition, they can take on "V" or "U" forms on the seismic profile. Deepwater canyon systems controlled the conveyance and gathering of sediments from land to sea and developed active gravity flow. The formation of submarine canyons is controlled by many factors [89,90], including sediment supply, change in sea level, regional structure activities, diapiric folds, and others. However, mass instability and erosion along the continental shelf–continental slope boundary are generally considered the major controlling factors [50,91]. In the research area, two deepwater channels developed successively. They are mainly lateral accretion and filling, and generally they have a lentoid appearance with a flat top and convex bottom. The internal structure is a medium wave amplitude diagonal type or prograde structure, showing the lateral migration characteristics during canyon filling (Figures 18 and 19). The oriented southeastward migration of the canyon might be related to the bottom flow action [92,93]. Its development scale became gradually small, its trenching depth became gradually small, and its angle became gradually gentle, evolving from lateral accretion to vertical accretion gradually.

The evolution of slump blocks is typically closely related to the geological environment and dynamic processes of deepwater canyons, and the accumulation of gas hydrates is also a result of interactions with these geological processes. Most of the filling material inside the eroded channel consists of deep-sea sediments that were deposited during the stationary phase of the gravity flow. The gravity flow eroded the channel wall over the long term, which caused the instability of the channel wall and slumping [50]. The slump block within the canyon ridge forms a gas hydrate stability zone that is conducive to gas hydrate accumulation (Figures 18 and 19). Inside the canyon and channel, there are usually no obvious BSRs. The sedimentary environment in deepwater canyon regions is relatively complex, accompanied by erosional processes and multiphase deposition. During the evolutionary process, slumping events frequently occur, and these slump blocks alter the region's topography and sedimentary environment, affecting the porosity, sand layer distribution, and connectivity of the reservoir [86,94]. The sediments surrounding the canyon have good sealing properties, providing a barrier for the stable accumulation of gas hydrates.

5.3. Migration and Accumulation Mechanism of Gas Hydrate

The research area lies in the northern continental slope of the South China Sea, with a water depth exceeding 1000 m, belonging to the deepwater areas. Judging from its cause of formation, the action mechanism of sediment gravity currents (landslide, slumping, mudflow, grain flow, turbidity current, and mud flow) and bottom flow are its dominant sediment-driving mechanisms [2,95,96]. These differ greatly from the slow and continual tractive current sediment in shallow-water areas. The deepwater environment is characterized by mass flow, various bottom flows, and pelagic/hemipelagic deposit. Turbidity current is rare in canyons, but is common in basins. The bottom flow following the fathom line and the circulatory motion of wind-driven bottom flow are the important mechanisms for forming the deepwater environment outside the canyon.

The research area has the typical slumping-canyon erosion channel–front-end deepwater fan source–ditch–convergence coupling sedimentary system (Figure 20) [18,48,49]. The deepwater fan is composed of the upper fan, middle fan, and lower fan. The upper fan is mainly composed of the slope angle, the main canal with a natural barrier, and its complex, and the flat and low areas are on both sides of the main canal. The proximal end of the middle fan develops at the place below the slope break near the slope angle; there is a large area of mudflow sediment, and the eroded channel–lobe complex is its major sedimentary system. The far end of the middle fan develops a highly bent channel complex, highly bent channel–lobe complex, and wash-over fan and crevasse splay. There are slip blocks in the upper fan and middle fan. The lower fan is located at the terminus of the fan and develops a large area of lobes, mainly including the end lobes and end branch channel of the lower fan channel. The lower fan also developed bathyal argillaceous sediment in the deepwater sediment [50].

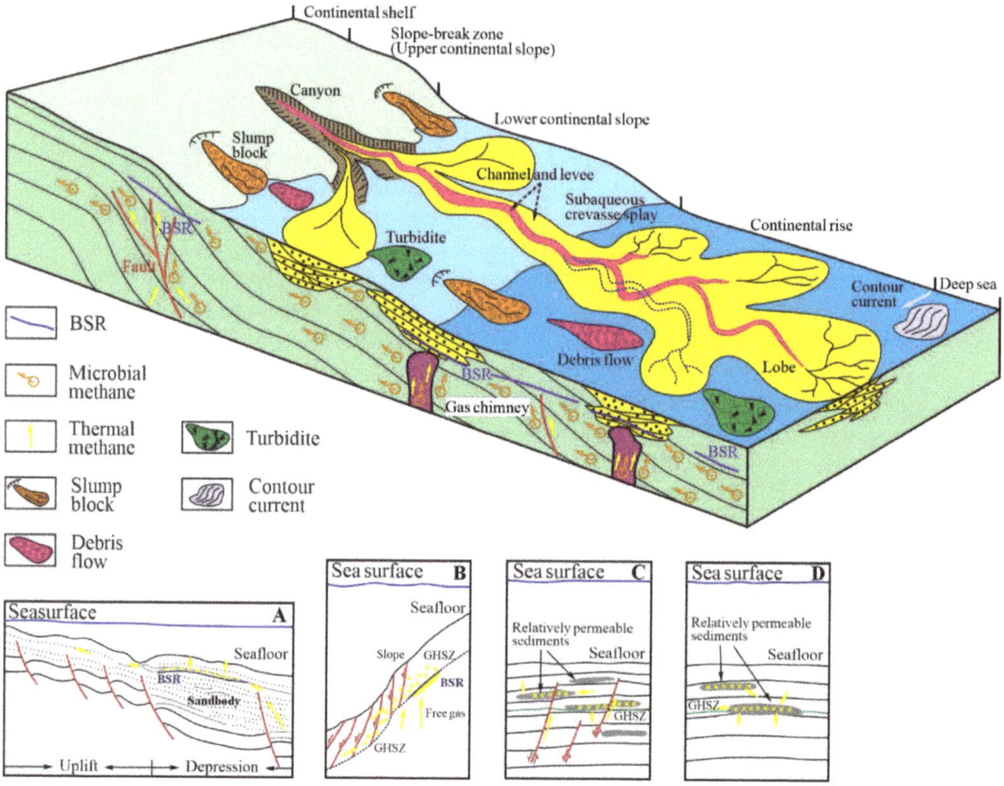

Figure 20. Ideal gas hydrate sedimentation and formation mode in the Shenhu drilling area in the northern continental slope of the South China Sea. The lower part of the figure is four types of hydrate accumulations. (**A**). Structural accumulation associated with a fault system, (**B**). Stratigraphic accumulation, (**C**). Control of gas hydrate in relative permeable layer by deep faults, (**D**). Gas hydrate in relative permeable layers. Arrows show fluid migration. Not to scale.

BSRs are widely distributed within the slumping-canyon erosion channel–deepwater fan sedimentary system. Their distribution is closely associated with the slumping canyon, characterized by high deposition rates, structural slope breaks, deepwater gravity flows, and contour current deposits. Most BSR distribution areas are in the seaward areas of steep relief variations, large topographical relief, convergence of long-term successive uplift, and sedimentation. The rapid loss of sediment, especially the gravity flow sediment under the slope, including sliding mass and slumped mass, are naceous mudflow and turbidity current, which can provide a good sediment reservoir bed and sealing conditions for the occurrence of natural gas hydrate [97,98]. Furthermore, in these BSR distribution areas, tectonic activity is strong. As a result, these areas can easily develop geologic bodies like fractures and diapiric folds, which can serve as passages for deep gas sources to move upward. The sediment sand bodies can be both passages for parallel migration of shallow layer air sources and the advantageous facies belt bearing the gas hydrate.

There are four main migration and accumulation mechanisms of gas hydrate:

A, Structural accumulation associated with a fault system: The structural slope break zone promotes the formation and accumulation of gas hydrate. The deep fault acts as a conduit for the upward migration of deep gas. This zone also features thickened sandstone layers, which facilitate lateral gas migration. The steep gradient and increasing water depth create a dynamic temperature and pressure environment, enabling free gas to migrate and form hydrate under the appropriate conditions (Figure 14) [99].

B, Stratigraphic accumulation: The strata stacking resulting from multiple stages of normal faulting can create an effective migration pathway and favorable reservoir conditions for gas hydrates. This process is closely associated with slump blocks, where the accumulation of fault-controlled strata contributes to the development of high-potential storage environments. Deep free gas migrates upward along the faults, utilizing these pathways to reach the surface. As the gas ascends, it encounters a suitable temperature and pressure regime, resulting in the formation of a gas hydrate stability zone (GHSZ) at the base of the slump block (Figure 17) [100].

C, Control of gas hydrate in relative permeable layer by deep faults: The deep free gas migrates upward along deep-seated faults, utilizing these pathways to travel towards relatively high-permeability formations. As the gas moves through these formations, it encounters favorable pressure and temperature conditions that allow it to transition into a gas hydrate stable zone. This process is akin to the formation of lithologic gas hydrate deposits, where gas is trapped within a specific lithology under stable environmental conditions (Figure 16) [101,102].

D, Gas hydrate in relative permeable layers: In the absence of faults, deep free gas migrates upward through high-permeability strata, which provide natural pathways for the gas to travel. As the gas rises through these permeable formations, it encounters suitable temperature and pressure conditions that facilitate its conversion into a gas hydrate stable zone. This process is similar to the formation of lithologic gas hydrate deposits, where gas accumulates in specific lithologies under stable environmental conditions (Figure 16) [101,103].

5.4. *Implications for Sedimentation and Controls on Gas Hydrate Distribution*

Sedimentation plays a key role in gas hydrate formation. Fine-grained sediments, like silts and clays, create conditions favorable for hydrate formation, especially in methane-rich environments [5,14]. The rate of sedimentation influences the thickness of the hydrate stability zone (HSZ) and the available pore space for hydrate accumulation [104,105]. In areas with rapid sedimentation, such as deepwater slope regions, free gas can be transported upward, promoting the formation of the hydrate stability zone [102,103]. Depositional facies, such as turbidites, create pathways for methane migration, affecting hydrate development. Sandy layers and interbedded turbidites facilitate methane movement, creating localized hydrate deposits [106,107]. Conversely, in areas with slower sedimentation, methane may not reach the necessary depth for hydrate formation, leading to limited or no hydrate presence.

Structural features like gas chimneys, faults, and deepwater channels are crucial in controlling methane migration and hydrate accumulation. Gas chimneys act as vertical conduits for methane, and, when they intersect the HSZ, they can lead to hydrate accumulation [83,108]. Faults and fractures also serve as vertical pathways for gas, and their intersection with sedimentary layers can create localized hydrate reservoirs [109]. Deepwater channels, such as unidirectional migrating channels, influence hydrate distribution by controlling slump occurrences. These channels affect sedimentary structures and erosion, which in turn impact the spatial distribution of gas hydrates [86,110].

Understanding sedimentary processes and geological controls on hydrate distribution is essential for exploration and resource assessment. By analyzing depositional environments, such as deepwater fans, submarine canyons, and slope basins, along with structural elements like gas chimneys and faults, researchers can identify areas with high potential for gas hydrate accumulation. Combining this data with geophysical information, such as seismic surveys, can improve predictions of hydrate distribution and enhance resource assessments, supporting future exploration efforts.

6. Conclusions

(1) BSRs in the hydrate drilling area of the Shenhu slope are mainly distributed in the continental slope ridge and slip blocks of the side slope; they are closely related to the development of the gas chimney. Most are at depths of 1200–1350 m in strata sequence IV and strata sequence III. Strata sequence IV has four BSR distribution areas, mainly in the northwest and northeast of the research area, and the BSRs vary widely in size. Strata sequence III only has one BSR distribution area; it lies in the southeast of the research area, with good continuity distributing in the form of ribbons. The drilling results show that hydrate develops in the bottom thin layer of the gas hydrate stability zone of strata sequence IV;

(2) The RMS amplitude attribute is sensitive to the flection of gas hydrates. The sand factor can be acquired by combining the RMS amplitude seismic attributes extracted and the drilling and logging results. The results show that the sand factor is generally smaller than 10% in each strata sequence, and that BSR disperses where the sand factor is relatively high (4–10%). The lithologic character of the hydrate layer mainly finds expression in silt and silty mud. The probability curve of granularity is mainly reflected by the suspension population with a lower slope. The porosity correlates positively with the hydrate saturation;

(3) The distribution of gas hydrates is governed by complex interactions between sedimentation, structural geology, and environmental factors. Submarine landslides and deepwater canyons play a crucial role in this process by altering sedimentary environments, increasing sand content, and enhancing reservoir formation. Slump blocks created by mass flow provide favorable conditions for hydrate accumulation, while deepwater canyons facilitate gas migration and accumulation through erosion and sediment deposition. Integrating these factors improves our understanding of gas hydrate formation and distribution, which is essential for both scientific advancements and resource management in the context of global energy and climate challenges.

Author Contributions: Conceptualization, Y.H. and X.Y.; methodology, Y.H., X.Y. and C.F.; validation, Y.H. and C.F.; investigation, Y.H.; data curation, Y.H.; writing—original draft preparation, Y.H.; writing—review and editing, Y.H., X.Y. and C.F.; visualization, Y.H.; supervision, X.Y. and C.F. All authors have read and agreed to the published version of the manuscript.

Funding: This research was funded by the National Natural Science Foundation of China—Youth Science Fund 42402150.

Data Availability Statement: Data are contained within the article.

Acknowledgments: We sincerely appreciate the data support provided by GMGS.

Conflicts of Interest: The authors declare no conflicts of interest.

References

1. Matsumoto, R.; Ryu, B.; Lee, S.; Lin, S.; Wu, S.; Sain, K.; Pecher, I.; Riedel, M. Occurrence and exploration of gas hydrate in the marginal seas and continental margin of the Asia and Oceania region. *Mar. Pet. Geol.* **2011**, *28*, 1751–1767. [CrossRef]
2. Yu, X.; Wang, J.; Liang, J.; Li, S.; Zeng, X.; Li, W. Depositional characteristics and accumulation model of gas hydrates in northern South China Sea. *Mar. Pet. Geol.* **2014**, *56*, 74–86. [CrossRef]
3. Sain, K.; Gupta, H. Gas hydrates in India: Potential and development. *Gondwana Res.* **2012**, *22*, 645–657. [CrossRef]
4. Tréhu, A.M.; Long, P.E.; Torres, M.E.; Bohrmann, G.; Rack, F.R.; Collett, T.S.; Goldberg, D.S.; Milkov, A.V.; Riedel, M.; Schultheiss, P. Three-dimensional distribution of gas hydrate beneath southern Hydrate Ridge: Constraints from ODP Leg 204. *Earth Planet. Sci. Lett.* **2004**, *222*, 845–862. [CrossRef]

5. Su, M.; Luo, K.W.; Fang, Y.X.; Kuang, Z.G.; Yang, C.Z.; Liang, J.Q.; Liang, C.; Chen, H.; Lin, Z.X.; Wang, C.; et al. Grain-size characteristics of fine-grained sediments and association with gas hydrate saturation in Shenhu Area, northern South China Sea. *Ore Geol. Rev.* **2021**, *129*, 103889. [CrossRef]
6. Gu, Y.H.; Li, S.J.; Song, Z.Y.; Lu, H.F.; Xu, C.L.; Sun, J.X.; Wang, Y.; Li, X.S.; Linga, P.; Yin, Z.Y. Analysis on a five-spot well for enhancing energy recovery from silty natural gas hydrate deposits in the South China Sea. *Appl. Energy* **2024**, *376*, 124237. [CrossRef]
7. Wang, D.; Ning, F.; Lu, J.; Lu, H.; Kang, D.; Xie, Y.; Li, J.; Sun, J.; Ou, W.; Liu, Z. Reservoir characteristics and critical influencing factors on gas hydrate accumulations in the Shenhu area, South China Sea. *Mar. Pet. Geol.* **2021**, *133*, 105238. [CrossRef]
8. Huang, L.; Yin, Z.; Wan, Y.; Sun, J.; Wu, N.; Veluswamy, H.P. Evaluation and comparison of gas production potential of the typical four gas hydrate deposits in Shenhu area, South China sea. *Energy* **2020**, *204*, 117955. [CrossRef]
9. Qin, X.; Liang, Q.; Ye, J.; Yang, L.; Qiu, H.; Xie, W.; Liang, J.; Lu, C.; Lu, H.; Ma, B. The response of temperature and pressure of hydrate reservoirs in the first gas hydrate production test in South China Sea. *Appl. Energy* **2020**, *278*, 115649. [CrossRef]
10. Veluswamy, H.P.; Kumar, S.; Kumar, R.; Rangsunvigit, P.; Linga, P. Enhanced clathrate hydrate formation kinetics at near ambient temperatures and moderate pressures: Application to natural gas storage. *Fuel* **2016**, *182*, 907–919. [CrossRef]
11. Shou-Ji, L.; Xin, S.; Gao, H.; Qian, D.; Wei-Hai, C.; Beng-Kang, W.; Yong-Gong, L.; Qing-Hai, L. Geological controlling factors of gas hydrate occurrence in Qilian Mountain permafrost, China. *Earth Sci. Front.* **2013**, *20*, 223.
12. Zang, X.; Liang, D.; Wu, N. Gas hydrate formation in fine sand. *Sci. China Earth Sci.* **2013**, *56*, 549–556. [CrossRef]
13. Su, M.; Zhang, C.; Xie, X.; Wang, Z.; Jiang, T.; He, Y.; Zhang, C. Controlling factors on the submarine canyon system: A case study of the Central Canyon System in the Qiongdongnan Basin, northern South China Sea. *Sci. China Earth Sci.* **2014**, *57*, 2457–2468. [CrossRef]
14. Bai, H.X.; Pecher, I.A.; Adam, L.; Field, B. Possible link between weak bottom simulating reflections and gas hydrate systems in fractures and macropores of fine-grained sediments: Results from the Hikurangi Margin, New Zealand. *Mar. Pet. Geol.* **2016**, *71*, 225–237. [CrossRef]
15. Chen, F.; Su, X.; Zhou, Y. Late Miocene-Pleistocene calcareous nannofossil biostratigraphy of Shenhu gas hydrate drilling area in the South China Sea and variations in sedimentation rates. *Earth Sci.* **2013**, *38*, 1–9.
16. Wang, X.; Hutchinson, D.R.; Wu, S.; Yang, S.; Guo, Y. Elevated gas hydrate saturation within silt and silty clay sediments in the Shenhu area, South China Sea. *J. Geophys. Res. Solid Earth* **2011**, *116*, B05102. [CrossRef]
17. Sun, Y.; Wu, S.; Dong, D.; Ludmann, T.; Gong, Y. Gas hydrates associated with gas chimneys in fine-grained sediments of the northern South China Sea. *Mar. Geol.* **2012**, *311*, 32–40. [CrossRef]
18. Su, M.; Lin, Z.X.; Wang, C.; Kuang, Z.G.; Liang, J.Q.; Chen, H.; Liu, S.; Zhang, B.D.; Luo, K.W.; Huang, S.Q.; et al. Geomorphologic and infilling characteristics of the slope-confined submarine canyons in the Pearl River Mouth Basin, northern South China Sea. *Mar. Geol.* **2020**, *424*, 106–166. [CrossRef]
19. Dai, J.X. *Giant Coal-Derived Gas Fields and Their Gas Sources in China*; Academic Press: Cambridge, MA, USA, 2016; pp. 449–453.
20. Wang, C.; Zeng, J.H.; Zhang, Z.T.; Shi, N.; Lao, M.J.; Zhao, Q.; Dai, J.H.; Wang, F.F.; Liu, X.F. Origin and distribution of natural gas and oil in the Baiyun Depression, Pearl River Mouth Basin, South China Sea. *J. Pet. Sci. Eng.* **2018**, *170*, 467–475. [CrossRef]
21. Long, Z.L.; Peng, G.R.; Chen, C.; Wang, X.D.; Huang, Y.P.; Shi, C. Gas generation and main gas source rocks in Baiyun Sag and the Pearl River Mouth Basin of the northern South China Sea. *Nat. Gas Ind. B* **2024**, *11*, 661–678. [CrossRef]
22. Kuang, Z.; Guo, Y. The Sedimentary Fades and Gas Hydrate Accumulation Models Since Neogene of Shenhu Sea Area, Northern South China Sea. *Earth Sci.* **2011**, *36*, 914–920.
23. Lu, X.L.; Li, M.J.; Li, Y.C.; Yang, Y.C.; Yang, H.Z.; Wang, N.; Ran, Z.C.; Ali, S. Early Oligocene marine transgression and organic matter accumulation recorded in the upper Eocene to lower Oligocene Enping Formation of the Baiyun Sag, Pearl River Mouth Basin, South China Sea. *J. Asian Earth Sci.* **2024**, *259*, 105909. [CrossRef]
24. Yu, X.H.; Wang, J.Z.; Li, S.L.; Fang, J.N.; Jiang, L.Y.; Cong, X.R.; Liang, J.Q.; Sha, Z.B. The Relationship between Tectonic Subsidence and BSR of Upper Neogene in the Deep-Water Area of the Northern Continental Slope, South China Sea. *Acta Geol. Sin. Engl.* **2013**, *87*, 804–818. [CrossRef]
25. Foschi, M.; Etiope, G.; Cartwright, J.A. Seismic evidence of extensive microbial gas migration and trapping in submarine gas hydrates (Rakhine Basin, Bay of Bengal). *Mar. Pet. Geol.* **2023**, *149*, 106100. [CrossRef]
26. Sun, L.; Wang, X.; He, M.; Jin, J.; Li, J.; Yuanping, L.; Zhu, Z.; Zhang, G. Thermogenic gas controls high saturation gas hydrate distribution in the Pearl River Mouth Basin: Evidence from numerical modeling and seismic anomalies. *Ore Geol. Rev.* **2020**, *127*, 103846. [CrossRef]
27. Hsiung, K.; Saito, S.; Kanamatsu, T.; Sanada, Y.; Yamada, Y.; Team, N.E.J.S. Regional stratigraphic framework and gas hydrate occurrence offshore eastern India: Core-log-seismic integration of National Gas Hydrate Program Expedition 02 (NGHP-02) Area-B drill sites. *Mar. Pet. Geol.* **2019**, *108*, 206–215. [CrossRef]
28. Zheng, X.B.; Zhu, H.T.; Liu, Q.H. Seismic geomorphology applied to sediment dispersal patterns and an analysis of the stages of channel-lobe systems in the Yinggehai Basin, northern South China Sea. *Mar. Pet. Geol.* **2022**, *139*, 105602. [CrossRef]

29. Guo, C.F.; Tang, Y.; Fang, Y.X.; Wang, C.Y.; Wei, X.D.; Li, H.; Chao, P.; Yang, T.Y.; Song, Z.B.; Ren, J.Y.; et al. Magmatism along the Nansha Trough on the southern continental margin of the South China Sea: Recent evidence from along-strike seismic profile *Mar. Geol.* **2024**, *475*, 107344. [CrossRef]
30. Mitchum, R., Jr.; Vail, P.R.; Thompson, S., III. Seismic stratigraphy and global changes of sea level: Part 2. The depositional sequence as a basic unit for stratigraphic analysis: Section 2. Application of seismic reflection configuration to stratigraphic interpretation. In *M 26: Seismic Stratigraphy–Applications to Hydrocarbon Exploration*; AAPG Datapages, Inc.: Tulsa, OK, USA, 1977
31. Allawi, R.H.; Al-Mudhafar, W.J.; Thanh, H.V. Developing A Semi-Analytical Model for Estimating Mechanical Properties of Sandstone Reservoirs: Enhancing Applications in hydrocarbon production and Underground Gas Storage. *Geoenergy Sci. Eng.* **2024**, *240*, 213014. [CrossRef]
32. Catuneanu, O.; Galloway, W.E.; Kendall, C.G.S.C.; Miall, A.D.; Posamentier, H.W.; Strasser, A.; Tucker, M.E. Sequence stratigraphy: Methodology and nomenclature. *Newslett. Strat.* **2011**, *44*, 173–245. [CrossRef]
33. Shipley, T.H.; Houston, M.H.; Buffler, R.T.; Shaub, F.J.; Mcmillen, K.J.; LAOD, J.W.; Worzel, J.L. Seismic Evidence for Widespread Possible Gas Hydrate Horizons on Continental Slopes and Rises. *AAPG Bull.* **1979**, *63*, 2204–2213.
34. Saeki, T.; Fujii, T.; Inamori, T.; Kobayashi, T.; Hayashi, M.; Nagakubo, S.; Takano, O. Extraction of Methane Hydrate Concentrated Zone for Resource Assessment in the Eastern Nankai Trough, Japan. In Proceedings of the Offshore Technology Conference, Houston, TX, USA, 5–8 May 2008.
35. Liu, Y.; Liu, X. Applications of geophysical techniques to gas hydrate prediction. *Pet. Explor. Dev.* **2007**, *34*, 566.
36. Monteleone, V.; Minshull, T.A.; Marin-Moreno, H. Seismic characterisation of multiple BSRs in the Eastern Black Sea Basin. *Mar. Pet. Geol.* **2024**, *161*, 106604. [CrossRef]
37. Le, A.N.; Huuse, M.; Redfern, J.; Gawthorpe, R.L.; Irving, D. Seismic characterization of a Bottom Simulating Reflection (BSR) and plumbing system of the Cameroon margin, offshore West Africa. *Mar. Pet. Geol.* **2015**, *68*, 629–647. [CrossRef]
38. Bashir, Y.; Kemerli, B.D.; Yilmaz, T.; Saral, M.; GoKnar, E.C.; Korkmaz, E. Reconstruction of subsurface potential hydrocarbon reservoirs through 3D seismic automatic interpretation and attribute analysis. *Phys. Chem. Earth* **2024**, *136*, 103751. [CrossRef]
39. Li, W.; Yu, X.; Zeng, X.; Wang, J.; Shan, X. Study of Neogene sesimic and sedimentary facies in the hydrate survey area of Shenhu Region on the north margin of South China Sea. *Mar. Geol. Front.* **2013**, *29*, 18–25.
40. Lutome, M.S. Seismic geomorphology and evolution of Mid-Late Miocene deepwater channels offshore Taranaki basin, New Zealand. *Mar. Pet. Geol.* **2025**, *171*, 107202. [CrossRef]
41. Wang, G.X.; Wu, W.; Li, Q.; Liu, W.Q.; Zhou, Y.S.; Liang, S.Q.; Sui, Y.P. Seismic geomorphology and reservoir conditions of a Middle Miocene submarine channel system in the Taranaki Basin, New Zealand. *Mar. Pet. Geol.* **2024**, *172*, 107219. [CrossRef]
42. Kiswaka, E.B.; Harishidayat, D.; Mkinga, O.J.; Saroni, J.J. Late Cenozoic mass transport deposits in the offshore Tanzania continental margin. *J. Afr. Earth Sci.* **2024**, *218*, 105377. [CrossRef]
43. Wan, L.; Hurter, S.; Bianchi, V.; Li, P.; Wang, J.; Salles, T. The roles and seismic expressions of turbidites and mass transport deposits using stratigraphic forward modeling and seismic forward modeling. *J. Asian Earth Sci.* **2022**, *232*, 105110. [CrossRef]
44. Lin, Z.X.; Su, M.; Zhuo, H.T.; Chen, H.; Jin, J.P.; Huang, S.Q.; Wang, Z.N. Depositional architecture and evolution of Quaternary submarine canyon-fan system in the Baiyun Sag of the Pearl River Mouth Basin, northern South China Sea. *Mar. Pet. Geol.* **2024**, *170*, 107167. [CrossRef]
45. Jin, L.; Luan, X.W.; Raveendrasinghe, T.D.; Jiang, L.S.; Xue, Y.; Wei, X.Y.; Qiao, J.H.; Ma, H.Z. Influence of sedimentary processes and fault tectonics on the evolution of submarine canyons in the East Andaman Basin: Insights from high-resolution seismic data analysis. *Geomorphology* **2024**, *454*, 109179. [CrossRef]
46. Li, S.L.; Gong, C.L. Flow dynamics and sedimentation of lateral accretion packages in sinuous deep-water channels: A 3D seismic case study from the northwestern South China Sea margin. *J. Asian Earth Sci.* **2016**, *124*, 233–246. [CrossRef]
47. He, Y.L.; Kuang, Z.G.; Cheng, C.; Jiang, T.; Zhang, C.; Lu, B.Y.; Yang, C.Z.; Liu, J.Y.; Xiang, C.L. Effects of Depositional Processes in Submarine Canyons and Distribution of Gas Chimneys on Gas Hydrate Accumulation in the Shenhu Sea Area, Northern South China Sea. *Energies* **2022**, *16*, 234. [CrossRef]
48. Cai, Z.R.; Sui, X.; Yao, Y.J.; Yin, Z.X.; Yang, X.D.; Ju, D.; Zhu, R.W.; Wang, X.X.; Zhang, H.D. Preliminary analysis of the indicators and accumulation models of gas hydrates in the central Zhongjiannan Basin, western South China Sea. *Mar. Pet. Geol.* **2024**, *167*, 106942. [CrossRef]
49. Fu, C.; Li, S.L.; Yu, X.H.; Liang, J.Q.; Kuang, Z.G.; He, Y.L.; Jin, L.N. Patterns of gas hydrate accumulation in mass transport deposits related to canyon activity: Example from Shenhu drilling area in the South China Sea. *Acta Oceanol. Sin.* **2019**, *38*, 118–128. [CrossRef]
50. Li, L.; Wang, Y.; Xu, Q.; Zhao, J.; Li, D. Seismic geomorphology and main controls of deep-water gravity flow sedimentary process on the slope of the northern South China Sea. *Sci. China Earth Sci.* **2012**, *55*, 747–757. [CrossRef]
51. Bu, Q.; Xing, T.; Li, C.; Zhao, J.; Liu, C.; Wang, Z.; Zhao, W.; Kang, J.; Meng, Q.; Hu, G. Effect of Hydrate Microscopic Distribution on Acoustic Characteristics during Hydrate Dissociation: An Insight from Combined Acoustic-CT Detection Study. *J. Mar. Sci. Eng.* **2022**, *10*, 1089. [CrossRef]

52. Liu, T.; Bao, X.; Geng, J.; Zhu, X.; Li, A.; Tian, D. Estimation of Seismic Attenuation and Gas Hydrate Concentration from Surface Seismic Data at Hydrate Ridge, Cascadia Margin. *J. Geophys. Res. Solid Earth* **2024**, *129*, e2023JB027123. [CrossRef]
53. Passega, R. Texture as characteristic of clastic deposition. *AAPG Bull.* **1957**, *41*, 1952–1984.
54. Mulder, T.; Alexander, J. The physical character of subaqueous sedimentary density ows and their deposits. *Sedimentology* **2001**, *48*, 269–299. [CrossRef]
55. Wynn, R.B.; Piper, D.J.; Gee, M.J. Generation and migration of coarse-grained sediment waves in turbidity current channels and channel^lobe transition zones. *Mar. Geol.* **2002**, *192*, 59–78. [CrossRef]
56. Malinverno, A.; Goldberg, D.S. Testing short-range migration of microbial methane as a hydrate formation mechanism: Results from Andaman Sea and Kumano Basin drill sites and global implications. *Earth Planet. Sci. Lett.* **2015**, *422*, 105–114. [CrossRef]
57. Riedel, M.; Collett, T.; Malone, M.; Akiba, F.; Blanc-Valleron, M.; Ellis, M.; Guerin, G.; Hashimoto, Y.; Heuer, V.; Higashi, Y. Gas hydrate transect across northern Cascadia margin. *Eos Trans. Am. Geophys. Union* **2006**, *87*, 325–332. [CrossRef]
58. Su, P.; Wei, W.; Sun, Y.; Lu, Y.; Cheng, H.; Han, W.; Zhang, W.; Liang, J. Double bottom simulating reflectors and tentative interpretation with implications for the dynamic accumulation of gas hydrates in the northern slope of the Qiongdongnan Basin, South China Sea. *China Geol.* **2024**, *7*, 422–444. [CrossRef]
59. Yu, X.; Zhang, Z. Characteristics of Neogene depositional systems on the northern continental slope of the South China Sea and their relationships with gas hydrate. *Geol. China* **2005**, *32*, 470–476.
60. Dillon, W.P.; Danforth, W.; Hutchinson, D.; Drury, R.; Taylor, M.; Booth, J. Evidence for faulting related to dissociation of gas hydrate and release of methane off the southeastern United States. *Lond. Geol. Soc. Lond. Spec. Publ.* **1998**, *137*, 293–302. [CrossRef]
61. Gering, K.L. Simulations of methane hydrate phenomena over geologic timescales. Part I: E¡ect of sediment compaction rates on methane hydrate and free gas accumulations. *Earth Planet. Sci. Lett.* **2003**, *206*, 65–81. [CrossRef]
62. Zhou, J.L.; Wang, X.J.; Collett, T.S.; Li, S.Z.; Kuang, Z.G.; Lu, Y.T.; Deng, W.; Yan, W.C.; Qian, J.; Jin, J.P. Characterization of a complex sand-rich gas hydrate reservoir system in the Indian marine continental margin with downhole log and seismic data. *Mar. Pet. Geol.* **2023**, *155*, 106370. [CrossRef]
63. Winters, W.J.; Pecher, I.A.; Waite, W.F.; Mason, D.H. Physical properties and rock physics models of sediment containing natural and laboratory-formed methane gas hydrate. *Am. Miner.* **2004**, *89*, 1221–1227. [CrossRef]
64. Ramana, M.; Ramprasad, T.; Paropkari, A.; Borole, D.; Ramalingeswara Rao, B.; Karisiddaiah, S.; Desa, M.; Kocherla, M.; Joao, H.M.; Lokabharati, P.; et al. Multidisciplinary investigations exploring indicators of gas hydrate occurrence in the Krishna—Godavari Basin offshore, east coast of India. *Geo-Mar. Lett.* **2009**, *29*, 25–38. [CrossRef]
65. Bahk, J.J.; Kim, D.H.; Chun, J.H.; Son, B.K.; Kim, J.H.; Ryu, B.J.; Torres, M.E.; Riedel, M.; Schultheiss, P. Gas hydrate occurrences and their relation to host sediment properties: Results from second Ulleung Basin gas hydrate drilling expedition, East Sea. *Mar. Pet. Geol.* **2013**, *47*, 21–29. [CrossRef]
66. Shahnazar, S.; Hasan, N. Gas hydrate formation condition: Review on experimental and modeling approaches. *Fluid Phase Equilibria* **2014**, *379*, 72–85. [CrossRef]
67. Majumdar, U.; Miller, N.C.; Ruppel, C.D. Neural Net Detection of Seismic Features Related to Gas Hydrates and Free Gas Accumulations on the northern U.S. Atlantic margin. *Interpretation* **2022**, *10*, T785–T806. [CrossRef]
68. Li, H.Y.; Liu, J.D.; Qu, C.X.; Song, H.Y.; Zhuang, X.L. A method for calculating gas hydrate saturation by dual parameters of logging. *Front. Earth Sci.* **2022**, *10*, 986647. [CrossRef]
69. Guan, J.; Wang, M.; Zhang, W.; Wan, L.; Haeckel, M.; Wu, Q. Representative Dynamic Accumulation of Hydrate-Bearing Sediments in Gas Chimney System since 30 Kyr BP in the QiongDongNan Area, Northern South China Sea. *J. Mar. Sci. Eng.* **2024**, *12*, 834. [CrossRef]
70. Xu, H.; Li, L.; Shu, X.; Wen, P.; Zhang, B. The seismic reflecting characteristics of gas hydrate bearing strata and its possible distribution in the South China Sea. *Appl. Geophys.* **2006**, *3*, 42–47. [CrossRef]
71. Constant Agnissan, A.; Guimpier, C.; Terzariol, M.; Fandino, O.; Cheron, S.; Riboulot, V.; Desmedt, A.; Ruffine, L. Influence of Clay-Containing Sediments on Methane Hydrate Formation: Impacts on Kinetic Behavior and Gas Storage Capacity. *J. Geophys. Res. Solid Earth* **2023**, *128*, e2023JB027333. [CrossRef]
72. Gao, R.; Wang, W.; Xiong, X.; Li, J.; Xu, C. Effect of curing temperature on the mechanical properties and pore structure of cemented backfill materials with waste rock-tailings. *Constr. Build. Mater.* **2023**, *409*, 133850. [CrossRef]
73. Hu, X.; Zhu, X.M.; Jin, X.L.; Huang, C.; Cheng, C.L.; Xiu, J.L.; Ren, X.C.; Zheng, D. Seismic sedimentology analysis of meter-scale thin sand bodies in the Jurassic Qigu Formation, Yongjin area, Junggar Basin. *Mar. Pet. Geol.* **2024**, *169*, 107079. [CrossRef]
74. Luo, K.W.; Liu, S.; Su, M.; Zhuo, H.T.; Cui, H.Q.; Chen, H.; Lei, Y.P.; Yu, M.D. Geomorphology of a bended submarine canyon in Wanhu Seamount region, northern South China Sea: Insights from manned submersible observation and measurement. *Geomorphology* **2021**, *392*, 107902. [CrossRef]
75. Jiao, H.J.; Wu, C.H.; Rodriguez-Lopez, J.P.; Sun, X.M.; Yi, H.S. Late Cretaceous plateau deserts in the South China Block, and Quaternary analogues; sedimentology, dune reconstruction and wind-water interactions. *Mar. Pet. Geol.* **2020**, *120*, 104504. [CrossRef]

76. Wang, J.X.; Ji, Y.K.; Liu, C.L.; Ning, F.L.; Meng, Q.G.; Zhao, Y.P.; Li, J.; Zhang, Z.; Zhang, Y.C.; Cai, F. Dependence of the hydrate-based CO2 storage characteristics on sand particle size and clay content in unconsolidated sediments. *Chem. Eng. J.* **2024**, *501*, 157497. [CrossRef]
77. Chen, X.J.; Zhou, K.; Jiang, P.; Zhao, X.B.; Yao, G.I. Identification, characterization, and up-scaling of pore structure facies in the low permeability reservoirs: Insight into reservoir quality evaluation and sweet-spots analysis. *Mar. Pet. Geol.* **2024**, *162*, 106646. [CrossRef]
78. Xu, N.; Wu, S.G.; Shi, B.Q.; Lu, B.; Xue, L.Q.; Wang, X.J.; Jia, Y. Gas hydrate associated with mud diapirs in southern Okinawa Trough. *Mar. Pet. Geol.* **2009**, *26*, 1413–1418. [CrossRef]
79. Lei, C.; Ren, J.Y.; Clift, P.D.; Wang, Z.F.; Li, X.S.; Tong, C.X. The structure and formation of diapirs in the YinggehaieSong Hong Basin, South China Sea. *Mar. Pet. Geol.* **2011**, *28*, 980–991. [CrossRef]
80. Ren, J.F.; Cheng, C.; Jiang, T.; Kuang, Z.G.; Lai, H.F.; Liang, J.Q.; Chen, Z.G.; Li, T. Faults and gas chimneys jointly dominate the gas hydrate accumulation in the Shenhu Area, northern South China Sea. *Front. Mar. Sci.* **2023**, *10*, 1254410. [CrossRef]
81. Brown, H.E.; Holbrook, W.S.; Hornbach, M.J.; Nealon, J. Slide structure and role of gas hydrate at the northern boundary of the Storegga Slide, offshore Norway. *Mar. Geol.* **2006**, *229*, 179–186. [CrossRef]
82. Fabre, M.; Loncke, L.; Riboulot, V.; Ker, S. Did gas-hydrate dissociation promote slope instability in the western Black Sea after the end of the last glacial period? *Mar. Pet. Geol.* **2024**, *170*, 107088. [CrossRef]
83. Pehlivanova, R.; Hristova, R. Gas chimneys in the northwestern part of the Black Sea Basin: Nature and seismic interpretation. *Rev. Bulg. Geol. Soc.* **2024**, *85*, 121–124. [CrossRef]
84. Jackson, B.A. Seismic evidence for gas hydrates in the North Makassar Basin, Indonesia. *Pet. Geosci.* **2004**, *10*, 227–238. [CrossRef]
85. Zyrianova, M.V.; Collett, T.S.; Boswell, R. Characterization of the Structural–Stratigraphic and Reservoir Controls on the Occurrence of Gas Hydrates in the Eileen Gas Hydrate Trend, Alaska North Slope. *J. Mar. Sci. Eng.* **2024**, *12*, 472. [CrossRef]
86. Lu, C.; Yao, Y.; Gong, Y.; Wu, S.; Li, X. Deepwater canyons reworked by bottom currents: Sedimentary evolution and genetic model. *J. Earth Sci. China* **2012**, *23*, 731–743. [CrossRef]
87. Jiang, T.; Zhang, Y.; Tang, S.; Zhang, D.; Zuo, Q.; Lin, W.; Wang, Y.; Sun, H.; Wang, B. CFD simulation on the generation of turbidites in deepwater areas: A case study of turbidity current processes in Qiongdongnan Basin, northern South China Sea. *Acta Oceanol. Sin.* **2014**, *33*, 127–137. [CrossRef]
88. Wang, X.X.; Sun, Q.L.; Wang, H.B.; Yin, S.R.; Wan, X.; Chen, J.; Hernandez-Molina, F.J. Flow conditions of the Quaternary Deep-water Current reconstructed by sediment waves in the northeastern South China Sea. *Mar. Geol.* **2024**, *477*, 107414. [CrossRef]
89. Lastras, G.; Arzola, R.G.; Masson, D.G.; Wynn, R.B.; Huvenne, V.A.; Huhnerbach, V.; Canals, M. Geomorphology and sedimentary features in the Central Portuguese submarine canyons, Western Iberian margin. *Geomorphology* **2009**, *103*, 310–329. [CrossRef]
90. Laursen, J.; Normark, W.R. Late Quaternary evolution of the San Antonio Submarine Canyon in the central Chile forearc (~33°S). *Mar. Geol.* **2002**, *188*, 365–390. [CrossRef]
91. Jobe, Z.R.; Lowe, D.R.; Uchytil, S.J. Two fundamentally different types of submarine canyons along the continental margin of Equatorial Guinea. *Mar. Pet. Geol.* **2011**, *28*, 843–860. [CrossRef]
92. Chen, H.; Xie, X.N.; Van Rooij, D.; Vandorpe, T.; Huang, L.; Guo, L.Y.; Su, M. Depositional characteristics and spatial distribution of deep-water sedimentary systems on the northwestern middle-lower slope of the Northwest Sub-Basin, South China Sea. *Mar. Geophys. Res.* **2013**, *34*, 239–257. [CrossRef]
93. Gong, C.L.; Wang, Y.M.; Zhu, W.L.; Li, W.G.; Xu, Q. Upper Miocene to Quaternary unidirectionally migrating deep-water channels in the Pearl River Mouth Basin, northern South China Sea. *AAPG Bull.* **2013**, *97*, 285–308. [CrossRef]
94. Gao, R.; Wang, W.; Zhou, K.; Zhao, Y.; Yang, C.; Ren, Q. Optimization of a Multiphase Mixed Flow Field in Backfill Slurry Preparation Based on Multiphase Flow Interaction. *Acs Omega* **2023**, *8*, 34698–34709. [CrossRef] [PubMed]
95. Shanmugam, G.; Shrivastava, S.; Das, B. Sandy debrites and tidalites of Pliocene reservoir sands in upper-slope canyon environments, offshore Krishna--Godavari Basin (India): Implications. *J. Sediment. Res.* **2009**, *79*, 736–756. [CrossRef]
96. Gao, R.; Zhou, K.; Zhou, Y.; Yang, C. Research on the fluid characteristics of cemented backfill pipeline transportation of mineral processing tailings. *Alex. Eng. J.* **2020**, *59*, 4409–4426. [CrossRef]
97. Pohl, F.; Eggenhuisen, J.T.; Tilston, M.; Cartigny, M. New flow relaxation mechanism explains scour fields at the end of submarine channels. *Nat. Commun.* **2019**, *10*, 4425. [CrossRef] [PubMed]
98. Bayet-Goll, A.; Sharafi, M.; Daraei, M.; Nasiri, Y. The influence of hybrid sediment gravity flows on distribution and composition of trace-fossil assemblages: Ordovician succession of the north-eastern Alborz Range of Iran. *Sedimentology* **2023**, *70*, 783–827. [CrossRef]
99. Wang, H.R.; Wang, Y.M.; Qiu, Y.; Peng, X.C.; Huang, Q.Z. Development and its tectonic activity's origin of turbidity current sediment wave in Manila Trench, the South China Sea. *Acta Sedimentol. Sin.* **2008**, *26*, 39.
100. Yang, M.Z.; Pan, A.D.; Sha, Z.B. Geological models of gas hydrates deposits along the continental margin. *Mar. Geol. Quat. Geol.* **2010**, *30*, 85–90.

101. Milkov, A.V.; Sassen, R. Economic geology of offshore gas hydrate accumulations and provinces. *Mar. Pet. Geol.* **2002**, *19*, 1–11. [CrossRef]
102. Liu, L.P.; Chu, F.Y.; Wu, N.Y.; Zhang, L.; Li, X.H.; Li, H.M.; Li, Z.G.; Zhang, W.Y.; Wang, X. Gas Sources, Migration, and Accumulation Systems: The Shallow Subsurface and Near-Seafloor Gas Hydrate Deposits. *Energies* **2022**, *15*, 6921. [CrossRef]
103. Ma, G.Z.; Zhan, L.S.; Lu, H.L.; Hou, G.T. Structures in Shallow Marine Sediments Associated with Gas and Fluid Migration. *J. Mar. Sci. Eng.* **2021**, *9*, 396. [CrossRef]
104. Lin, C.C.; Lin, A.; Tien, S.; Liu, C.S.; Horng, C.S.; Chen, G.Y.; Wang, Y. Canyon-infilling and gas hydrate occurrences in the frontal fold of the offshore accretionary wedge off southern Taiwan. *Mar. Geophys. Res.* **2014**, *35*, 21–35. [CrossRef]
105. Bernardes, L.; Carneiro, J.; Madureira, P.; Brandao, F.; Roque, C. Determination of Priority Study Areas for Coupling CO2 Storage and CH4 Gas Hydrates Recovery in the Portuguese Offshore Area. *Energies* **2015**, *8*, 10276–10292. [CrossRef]
106. Dirgantara, F.; Chiang, H.T.; Lin, A.T.S.; Liu, C.S.; Chen, S.C. Depositional influence of submarine channel migration on thermal properties of the Lower Fangliao Basin, offshore southwestern Taiwan. *Mar. Geophys. Res.* **2020**, *41*, 1. [CrossRef]
107. Ketzer, J.M.; Augustin, A.; Rodrigues, L.F.; Oliveira, R.; Praeg, D.; Pivel, M.A.G.; Dos Reis, A.T.; Silva, C.; Leonel, B. Gas seeps and gas hydrates in the Amazon deep-sea fan. *Geo-Mar. Lett.* **2018**, *38*, 429–438. [CrossRef]
108. Yang, R.; Su, M.; Qiao, S.H.; Cong, X.R.; Su, Z.; Liang, J.Q.; Wu, N.Y. Migration of methane associated with gas hydrates of the Shenhu Area, northern slope of South China Sea. *Mar. Geophys. Res.* **2015**, *36*, 253–261. [CrossRef]
109. Kutcherov, V.G.; Serovaiskii, A.Y. Contribution of Deep Hydrocarbons in Gas Hydrate Formation. *Chem. Technol. Fuels Oils* **2023**, *59*, 465–470. [CrossRef]
110. Meng, M.M.; Liang, J.Q.; Kuang, Z.G.; Ren, J.F.; He, Y.L.; Deng, W.; Gong, Y.H. Distribution Characteristics of Quaternary Channel Systems and Their Controlling Factors in the Qiongdongnan Basin, South China Sea. *Front. Earth Sci.* **2022**, *10*, 902517. [CrossRef]

processes

MDPI

Article

A Review of Potential Geological Hazards and Precautions in the Mining of Submarine Natural Gas Hydrate

Zhanghuang Ye [1], Wenqi Hu [2] and Qiang Yan [3,4,*]

1 School of Maritime Law and Transportation Management, Guangzhou Maritime University, Guangzhou 510725, China; chuckverna@sina.com
2 School of Tourism and Historical Culture, Jiangxi Science and Technology Normal University, Nanchang 330038, China; jxkjllxy@163.com
3 Chinese Academy of Geological Sciences, Beijing 100037, China
4 Research Center for Strategy of Global Mineral Resources, Chinese Academy of Geological Sciences, Beijing 100037, China
* Correspondence: cagsyq@163.com

Abstract: Natural gas hydrate (NGH hereafter), commonly known as combustible ice ($(CH_4)n{\cdot}mH_2O$), is an abundant non-conventional clean energy resource. It is mainly located in permafrost areas and submarine sediment layers at depths of 0–200 m and 300~3000 m underwater. Submarine NGH accounts for about 97%. Its commercial mining may be a solution to mankind's future energy problems, as well as the beginning of a series of geological risks. These risks can be divided into two categories: natural geological hazards and secondary geological accidents. Based on the viewpoints of Earth system science researchers, this paper discusses the main potential geo-hazards of submarine NGH mining: stratum subsidence, seafloor landslides, the greenhouse effect, sand piping, well blowout, and wellbore instability. To minimize the potential catastrophic impacts on the Earth's ecosystem or mechanical accidents, corresponding technical precautions and policy suggestions have been put forward. Hopefully, this paper will provide a useful reference for the commercial mining of NGH.

Keywords: NGH; geological risks; precautionary measures; commercial mining

1. Introduction

The massive use of fossil fuels is unsustainable and may have serious negative consequences. One example is climate change, which is becoming an increasingly significant concern for the international community. Petroleum consumption ascended to more than 97 million barrels per day in 2023. However, the proven global petroleum reserves left are about 1.7 trillion barrels, roughly enough for only 50 years [1]. In addition, the Russia–Ukraine conflict has profoundly affected the traditional energy supply chain. What is the future of human energy? The discovery of NGH has opened up a new solution to energy problems. NGH, commonly known as combustible ice ($(CH_4)n{\cdot}mH_2O$), is a crystalline solid compound formed by water molecules and methane molecules (with methane accounting for the majority) [2]. Widely located in submarine sediments and permafrost zones under conditions of low temperature and high pressure, it burns with a very small amount of carbon emission when it meets fire. NGH is mainly located in permafrost areas and within the top 0–200 m of submarine sediment layers at water depths of 300~3000 m. Submarine NGH accounts for about 97%. Its global reserves are estimated to be as high as 1.5×10^{15} m^3, twice the reserves of traditional fossil energy sources [3–5]. It is anticipated to be a feasible substitute for traditional fossil fuels, marking a shift to a sustainable-energy-dominated stage because of its cleanliness and large reserves [6–9]. Submarine NGH is mainly located in the 0–200 m sediment layer of the seabed surface, the so-called NGH Stability Zone (NGHSZ), where methane molecules are firmly fixed in the center of a cagelike structure in the stable form of NGH [10,11]. However, once the NGHSZ is destroyed due to submarine drilling, NGH will decompose into methane gas and water, and the volume will rapidly expand (1 unit of NGH dissociation can release about 160 times as much gas in the standard state) [12]. This phase transition will significantly reduce the bonding strength and shear strength of submarine sediments. Various geological hazards, such as stratum subsidence, seafloor landslides, the greenhouse effect, sand piping, well blowout, and wellbore instability, may be induced.

Depressurization [13,14], thermal stimulation [15], and inhibitor injection [16,17] are three mainstream technologies for gas production from submarine hydrate-bearing sediments. Most scholars believe that the commercial mining of NGH will have catastrophic impacts on the ocean ecosystem and pose potential threats to the subsequent production operation as depressurization or heating during mining will significantly change the NGH occurrence environment, characterized by low temperature and high pressure [18–24]. Submarine sediments generally exhibit poor cementation and diagenesis [11,25,26]. The long-term extraction of pore NGH will inevitably result in NGH-bearing sediments, and the overlying sediments will be subjected to deformation and subsidence of varying degrees [27–30]. He et al. [31] revealed that hydrate decomposition exacerbates the risk of wellbore instability, which will lead to more geological hazards, through the construction of coupled models. Zhang et al. [32] investigated the wellbore destabilization problem caused by deep-well exploitation and emphasized that deep-well exploitation may lead to chain reactions, such as seafloor landslides, stratum subsidence, and wellbore safety hazards. Li et al. [19] explored sediment deformation using the ABAQUS platform, showing that sediment instability around the wellbore occurs mainly in the early operation stage, which can lead to sand production. However, sediment instability near the seabed occurs in the late operation stage, which may lead to submarine landslides. Luo et al. [33] analyzed the deformation behavior of submarine sediments when depressurization and thermal stimulation techniques were applied to gas production. Song et al. [20] elaborated on the features and patterns of subsubmarine landslides induced by gas extraction from the seafloor. Wan et al. [27] analyzed the reservoir deformation associated with vertical wellbores. Wang et al. [34] experimentally studied the effect of hydrate dissociation on sediment deformation. Zhang et al. [35] revealed that the stability of hydrate reservoirs is worst in the direction of the minimum horizontal principal stress. Zhou et al. [36] investigated the mechanical responses of five selected elements at different locations in the hydrate gas production region, showing that the mechanical responses of hydrate-bearing sediments at specific locations are related to their hydrate dissociation status, which can be typically divided into the before, during, and after stages of hydrate dissociation. These studies collectively show that changes in the physical and mechanical properties of the strata triggered by mining activities can significantly affect the stability of the seabed geological structure through the interaction of multiple factors, and that the optimization of the mining process is required to control risk in engineering design.

Many countries have carried out NGH depressurization production testing (Table 1) [19,37–46]. It is high time we discussed the potential geological hazards and risk precautions for the NGH trial mining stage. In order to minimize potential risks, this paper will identify the potential geological hazards in submarine NGH mining and put forward corresponding precautions and warnings based on the theory of Earth system science.

Table 1. Summary of NGH mining by decompression methods.

NGH Mining Area	Time	Reservoir Characterization	Mode of Mining	The Total Gas Output of NGH
Nankai, Japan	March 2013	Water depth: 1000 m Burial depth: 300–360 m Type: Sand layer Average initial permeability: 20 mD	Depressurization production	Cumulative: 11.9×10^4 m^3 Average: 2.0×10^4 m^3/d
Nankai, Japan	May 2017	Water depth: 1000 m Burial depth: 300–360 m Type: Sand layer Average initial permeability: 20 mD	Depressurization production	Cumulative: 26.2×10^4 m^3 Average: 0.73×10^4 m^3/d
Shenhu Sea, China	July 2017	Water depth: 1266 m Burial depth: 203–277 m Type: Muddy chalk Average initial permeability: 2.9 mD	Depressurization production	Cumulative: 30.9×10^4 m^3 Average: 0.5×10^4 m^3/d
Shenhu Sea, China	April 2020	Water depth: 1225 m Burial depth: 207–253 m Type: Muddy chalk Average initial permeability: 2.38 mD	Depressurization production	Cumulative: 8614×10^4 m^3 Average: 2.87×10^4 m^3/d
Alaska, USA	May 2012	-	CO_2 displacement production	Cumulative: 30×10^4 m^3

2. Potential Geological Hazards in NGH Mining

In this paper, the potential geological hazards in NGH mining are divided into two main categories: natural geological hazards and secondary geological accidents. The former includes stratum subsidence, seafloor landslides, and the greenhouse effect; the latter includes sand piping, well blowout, and wellbore instability.

2.1. Stratum Subsidence

Stratum subsidence refers to the stratum vertical displacement resulting from crustal movement, industrial mining, or other factors. NGH mining will induce the decline of submarine strata. The faster the hydrate dissociation speed, the greater the stratum subsidence rate [47]. When NGH is extracted by drilling tools, hydrate dissociation is accelerated, stratum porosity is increased, and excessive pore pressure is generated [32]. The original shearing strength and mechanical equilibrium system of the seafloor are disrupted, which leads to vertical displacement of the sedimentary layer, and ultimately results in stratum subsidence (Figure 1a).

Li et al. [48] conducted a ten-year simulation experiment on gas hydrate extraction to deeply analyze the dynamic stratum subsidence. The experiment revealed the functional relationship between vertical displacement and well depth over the production times of 6 days, 30 days, 1 year, 5 years, and 10 years. The results show that the stratigraphic subsidence is particularly significant around the upper part of the wellbore, and the subsidence tendency becomes worse with time. However, the rate of subsidence slows significantly after 1 year of mining. Notably, a slight stratum uplift was observed near the lower part of the wellbore, which peaked at 4.2 cm at 30 days (Figure 1b). The study highlights that long-term mining-induced stratum subsidence may cause serious damage to offshore drilling facilities, and marine biological environments.

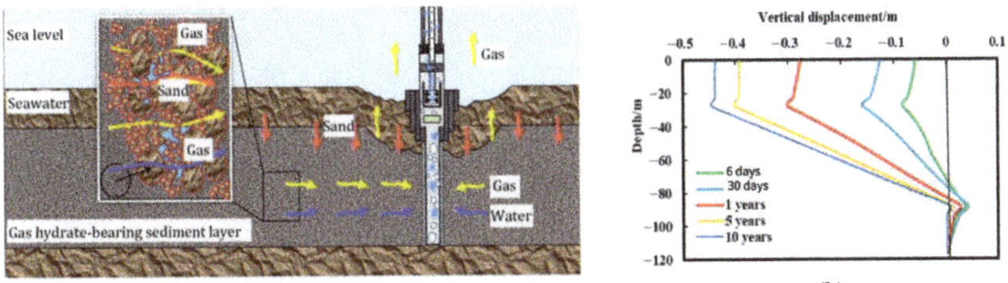

(a) (b)

Figure 1. Stratum subsidence due to NGH drilling (modified from reference [48,49]). (**a**) Diagram of NGH migration and stratum change. (**b**) Functional relationship between vertical displacement and borehole depth (reproduced with permission from [48]; [*Energy & Fuels*]; published by the American Chemical Society [2021]).

2.2. Seafloor Landslide

A seafloor landslide is a phenomenon where the seabed sediments collapse to form a flowing body under the influence of such natural factors as earthquakes or human factors, and the body moves along the deep sea under the force of gravity. According to the forms of sediment collapse, they can be classified as block landslides, fluvial landslides, or shallow slab landslides. According to the movement modes, they include creeping, sliding, turbidity flows, block flows, and debris avalanches. Five typical seafloor landslides have been recorded. Studies show that there is a strong coupling relationship between submarine landslides and hydrate decomposition in the South China Sea [50]. Ma et al. [51] have pointed out that seafloor landslides occurring during NGH exploitation pose a fatal threat to marine ecology and human safety. In addition, if the large amount of gas produced by hydrate decomposition is not properly released in time, the ultra-high pore pressure will further deteriorate the stability of seafloor slopes [52,53].

Both natural environmental factors and human factors can result in seafloor landslides in NGH-bearing sediment strata [54,55]. The natural environmental factors mainly include the NGH dissociation caused by global warming, and the human factors are mainly the temperature and pressure changes of the hydrate reservoir caused by NGH mining. No matter what factors they are, it all comes down to the basic incentive cause: submarine NGH dissociation. Hydrate extraction leads to the instability of the lower boundary of NGH, further movement of large block of sediments under the force of gravity currents, and seafloor landslides happen (Figure 2). With the advancement of the commercial mining of NGH in the future, the possibility of triggering seafloor landslide will increase. Seafloor landslides will impair the operation of seafloor monitoring tools and drilling equipment.

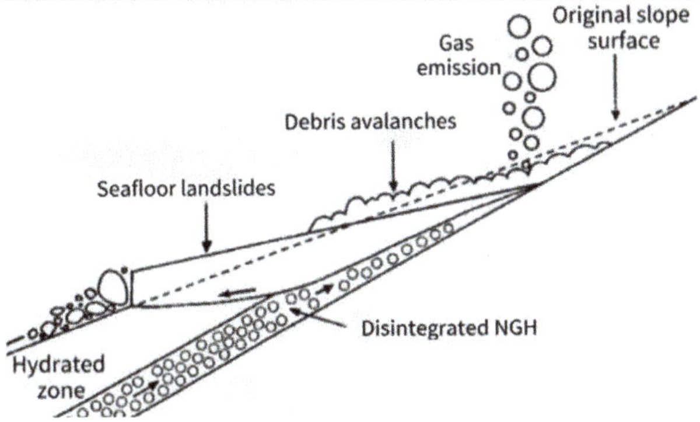

Figure 2. Seafloor landslides resulting from NGH extraction (modified from reference [56]).

2.3. Greenhouse Effect

The greenhouse effect will lead to significant climate change and extreme natural disasters, such as tsunamis, droughts, and floods, a growing global concern for international communities. On the one hand, in situ emission of natural gas was observed at the seafloor of the Black Sea [57]. On the other hand, geological hazards such as submarine earthquakes or volcanoes can create fractures or faults on the seafloor surface, resulting in a sudden and plentiful release of pore pressure gas. In addition, submarine NGH extraction will emit more greenhouse gases and eventually intensify the greenhouse effect. The marine temperature simulation in the Sea of Okhotsk in 2014, the Gulf Stream, and the Gulf of Mexico showed a 1–2 °C increase in 100 years due to submarine NGH extraction (Figure 3).

Methane-derived carbon is absorbed by Planktonic Foraminifera, Benthic Foraminifera, and carbonate rocks after methane seepage. This process is reflected in the following two chemical equations: $CH_4 + O_2 \rightarrow H_2O + CO_2$ (1); and $CH_4 + SO_4^{2-} \rightarrow HS^- + HCO_3^-$ (2). The carbon values can be measured by δ13C in the biomarkers and carbonate rocks. During geological history, NGH emission resulted in the end of the 'snowball' Earth during the Neoproterozoic era and a temperature increase of 6–12 °C in the Northern Hemisphere during the Paleocene–Eocene [58,59]. Methane is more harmful in terms of global warming than carbon emissions in industry. IPCC (UN Intergovernmental Panel on Climate Change) studies show that the greenhouse effect derived from CH_4 is 82.5 times as high as that from CO_2 on a 20-year time scale, and 29.8 times as high on a 100-year time scale [60].

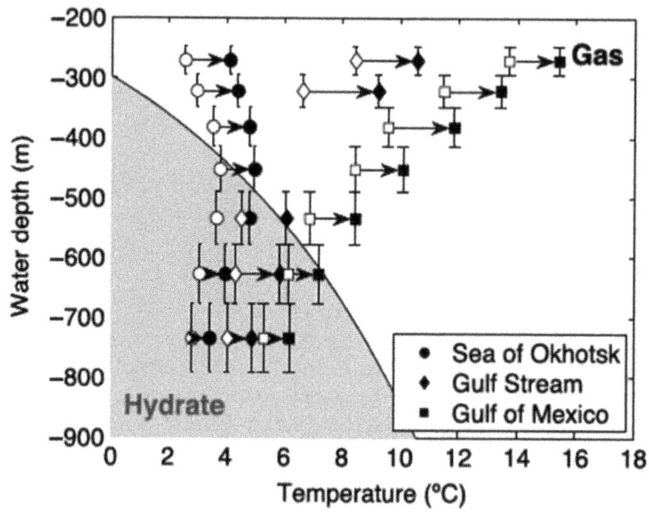

Figure 3. Marine temperature simulation showing a 1–2 °C increase (modified from reference [61]).

2.4. Sand Piping

Sand piping refers to the phenomenon where sand surges into the wellbore due to the pressure difference between the inside and the outside in the petroleum and gas drilling process, resulting in the blockage of the transportation passage. When the NGH dissociation (D_H) rate increases from 30% to 60%, the plastic strain increases from 2.3% to 2.8%. The probability is that sand piping will happen (Figure 4).

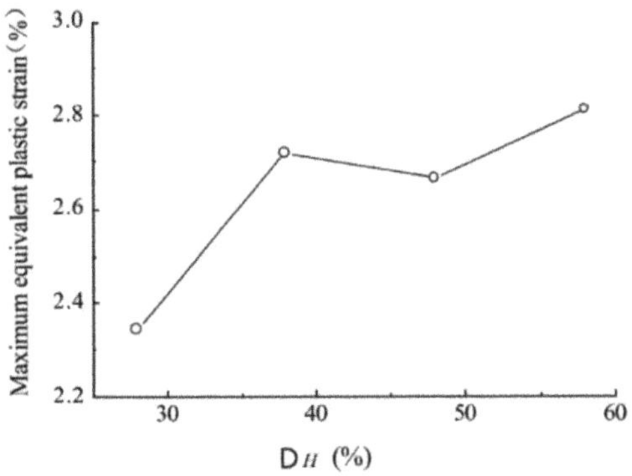

Figure 4. The relationship between NGH solubility and plastic strain ([62], reproduced with permission from [Yan, C. et al.]; [*Journal of natural gas science and engineering*]; published by ScienceDirect [2018]).

The commonly adopted depressurization method is more likely to result in a large amount of sand entering the wellbore and forming a blockage in mining. Excessive sand piping can cause serious damage to drilling tools, instability, and blockage of the wellbore. A typical example was when Japan failed in submarine NGH extraction due to sand piping.

2.5. Well Blowout

A well blowout is a phenomenon where petroleum, gas, or water suddenly erupts in an uncontrolled state due to pressure inconsistencies between the inside of the well and the outside. In offshore drilling operations, blowout accidents are extremely serious hazards, directly threatening physical safety, the marine environment, and economic benefits. In 2003, a well blowout occurred in Chongqing, with the tragic consequence of 243 deaths. NGH extraction also has the potential risk of well blowouts. When the pressure outside the well wall is too strong, excessive hydrate will surge into the well, which may lead to a blowout. Pipeline blocks, gas leakage, and even explosions may be induced when there is insufficient space inside the well (Figure 5).

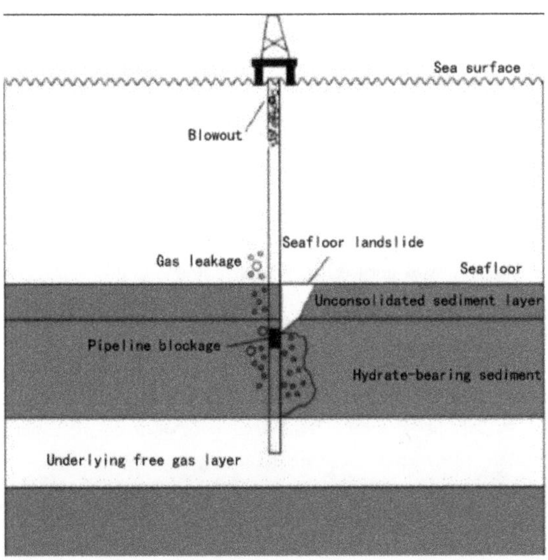

Figure 5. Principle of well blowout in NGH extraction (modified from reference [63]).

Regarding well blowouts formed during the NGH extraction process, scholars have mainly conducted research in three major forms: case study research, risk model construction, and experimental process simulation. Chen et al. [64] constructed a Bayesian network mode, specifically applied to offshore drilling operations, and simulated the blowout accident in the South China Sea. Johansen [65] conducted research on deep-water blowout cases and compared the research results with shallow-water blowout cases. Clarkson's deep-water oil and gas model simulated 30 blowout scenarios based on many actual cases. The simulation clearly showed significant differences in the prevalence of blowout plumes under different environmental conditions and different types of gases [66]. Sun et al. [67] simulated seven types of well blowouts by building a multinomial flow model and classified the blowouts into three phases: development of well kick, development of blowout, and a borehole stage with little mud.

2.6. Wellbore Instability

Wellbore instability is primarily a function of how submarine strata respond to the induced stress focused around the wellbore during various drilling activities. Many models have been developed for instability analysis and much progress has been made so far. However, wellbore instability continues to pose a considerable threat to successful well operations. A case in point is the rig explosion in the Gulf of Mexico on 20 April 2010, which was directly attributed to wellbore stability failure. The accident resulted in the deaths of 11 operators, serious contamination of more than 2000 km^2 of sea area, and great damage to the marine ecosystem. This incident highlights the disastrous consequences of wellbore instability in practical engineering under complex geological conditions.

Quantitative risk evaluations of the wellbore instability in high-temperature and high-pressure wells have been carried out by scholars [68–71]. More details related to wellbore instability are shown in Table 2 below [72,73].

Table 2. Types, conditions, examples, causes, and symptoms of wellbore instability.

Type	Occurrence Conditions	Typical Examples	Causes	Instability Symptoms
Tensile failure (stress failure)	When the strain created by borehole pressure exceeds the internal stress of the rock formation	Wellbore fracturing, hole enlargement	(1) Physical causes: tectonically stressed formations, anomalously high in situ stresses, naturally over-pressured collapse, naturally fractured or faulted formations (2) Man-made factors: drilling fluid density, well inclination, temporary wellbore pressures, drill string vibrations, drilling fluid temperature, induced over-pressured seafloor collapse, poor hole cleaning (3) Rock–fluid interaction	(1) Direct: oversized hole, undergauged hole, excessive volume of cuttings, excessive volume of cavings, cavings at the surface, hole fill after tripping, excess cement volume required, deviation control problems. (2) Indirect: high torque and drag, hanging up of drill string, casing, or coiled tubing, increased circulating pressures, stuck pipe, excessive drill string vibrations, drill string failure, inability to run logs, poor logging response, annular gas leakage due to poor cement job, keyhole seating, excessive doglegs
Shear failure	When the strain created by borehole pressure is less than the internal stress of the rock formation	Wellbore collapse, tight holes		

Maximizing the prevention of potential geological hazards is a crucial step in the future commercial mining of submarine NGH.

3. Technical Precautions

Previous studies [74–77] have provided the theoretical basis for the prevention and control of geological hazards in NGH extraction. This paper proposes the following technical precautions from the perspective of marine geological engineering.

3.1. Optimize the CO_2-EGR Method to Alleviate Stratum Subsidence

With the increase in gas production in NGH mining, the degree of stratum subsidence will increase. Our aim is to minimize the stratum subsidence (<0.05 m) while ensuring the maximum outputs. Through numerical simulation, Xin et al. [78] revealed that the stratum subsidence caused by NGH depressurization mining mainly occurs in the early stages. Therefore, the depressurization range should be reduced in the early mining stage. Based on a coupled geomechanical model, Lin et al. [79] simulated the subsidence of seafloor strata during NGH drilling, showing that significant improvements in seafloor subsidence can be achieved with the optimization of carbon-dioxide-enhanced gas recovery (CO_2-EGR). Two cases of pure depressurization and CO_2 injection depressurization were simulated by experiments, showing that the subsidence is only 6.8% of pure decompression when the pressure drop is 30%, that is, the degree of stratum subsidence is significantly reduced when carbon dioxide is injected earlier. The mechanism is that the injection rate of CO_2 is subject to double constraints: One is that the maximum injection rate is limited to 3900 standard scm/day, and the other is limited by the maximum bottom hole pressure. When the CO_2 injection rate reached the maximum, it was found to be a stable injection parameter that keeps a balance between seafloor subsidence and NGH production. However, the technology is still in the experimental stage and has not yet been tested in the field, and more efforts need to be made in further research.

3.2. Fill Foamed Cement to Control Seafloor Landslides

For the seafloor landslides caused by NGH extraction, currently there are four slurry cementing operations. The most effective method is to apply a foaming machine, initially used in the field of making thermal insulation material, to drilling technology. The voids in the seafloor strata produced in NGH extraction can be quickly filled. A cable winch on the drilling platform is used to control the descending depth, filling the foamed cement slurry through the shot hole while a steering gear underneath is used to regulate the drilling depth (Figure 6). The study shows that filling foamed cement slurry at the same time as drilling can effectively support the stratum and greatly avoid the occurrence of seafloor landslides [80].

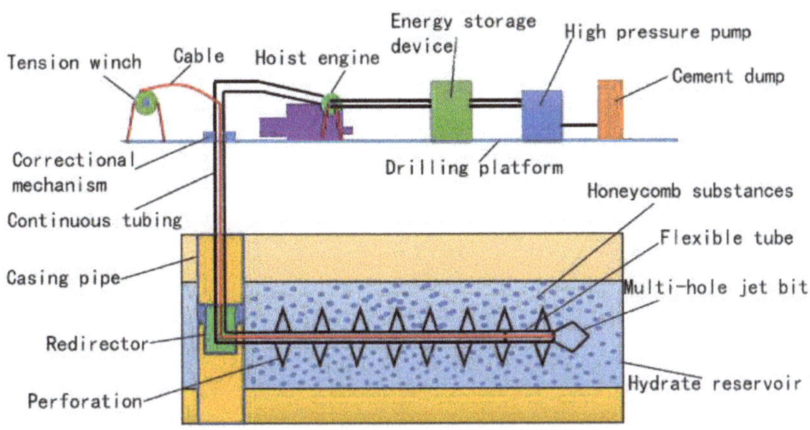

Figure 6. Principle of foaming machine in NGH drilling (modified from reference [80]).

3.3. Improve Drilling Technology to Mitigate the Greenhouse Effect

There are three main methods to slow down the emission of greenhouse gases: First, the use of petroleum-based or water-based drilling fluids bearing hydrate inhibitors can effectively depress the dissociation rate of NGH, ultimately slowing down the emission of greenhouse gases such as methane [81]. Second, the use of an emulsifier replacement can convert CH_4 produced in the dissociation of NGH into CO_2. As stated previously, CO_2 is a less harmful greenhouse gas than CH_4. Tian et al. [82] have shown that the CO_2

replacement method is a very efficient NGH mining method. When the recovery efficiency of CH_4 achieved by injecting the CO_2 mixture reaches about 6.95%, the storage rate of CO_2 is about 0.3. Third, biotechnology can absorb methane. Methanophilus, a kind of fungus which mainly feeds on methane, can be used to absorb the methane produced during hydrate dissociation. Our study shows that the second method will help promote a greener and cleaner gas hydrate extraction in the submarine environment.

3.4. Install a New Solid–Liquid Separator to Reduce Sand Piping

By installing an axial-flow annular in situ device, the target liquids are pumped out, leaving the unwanted solids on the seafloor during the mining process. There is no need for drilling tools to pump excessive sand-bearing material, reducing the sand piping (Figure 7).

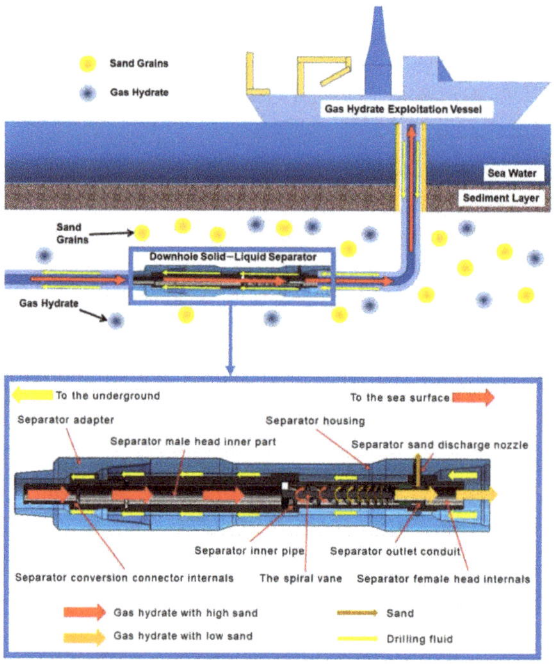

Figure 7. Principle of solid–liquid separator in NGH drilling ([83], reproduced with permission from Nie, Q. et al.; [*Energies*]; published by MDPI [2022]).

In addition, Figure 8 shows that sand production can be zero in theory if no pressure gradient is applied. However, this does not make any sense as there cannot be any gas production either. It is important to notice that the sand production is almost kept constant after gas production volumes of 1000 m^3. In other words, sand production is not affected by the rate of depressurization at the later stage. P8t2 has significantly lower sand production compared to the H5t2 stage. It is obvious that lowering the decompression rate is an effective operation strategy to reduce the early sand production stage.

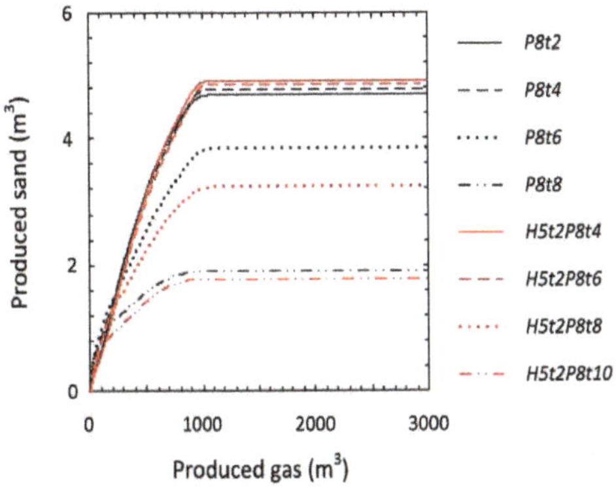

Figure 8. Sand production against gas production at different pressure gradients ([84], reproduced with permission from Uchida, S et al.; [*International Journal of Rock Mechanics and Mining Sciences*]; published by Science Direct [2016]).

3.5. Adopt New Inhibitor to Lower Well Blockage

At present, the issue of wellbore blockage is mainly alleviated by traditional inhibitors. Inhibitors can be divided into three main categories: thermodynamic inhibitors, mixed inhibitors, and kinetic hydrate inhibitors (Table 3) [85–97]. Yang et al. [89] have prepared a new low-dose polymer hydrate inhibitor HAY through monomers such as tetrahydrofuran and N-butenylpyrrolidone, which can more effectively inhibit hydrate formation and increase NGH production. Qin [90] found that the hydrate inhibition effects of self-made reagent PVP-BP are outstanding. The self-made reagent has no hydrate formation in the pipeline for up to 33 h at 8.4 K undercooling. When the temperature exceeds 8.4 K, hydrate formation is still less. Bu et al. [98] experimented on the effect of lecithin on mechanical properties at different concentrations, and confirmed that a 1.5% lecithin concentration in the cement slurry inhibits NGH decomposition. These studies provide a basis for optimizing the inhibitor formulation, suggesting that compounded inhibitors are more effective than single kinetic inhibitors.

Table 3. Types of, experiments on, and functions of NGH inhibitors.

Inhibitor Type	Experiment Method	Experimental Result	Inhibitor Mechanism	Inhibitor Effect
Thermodynamic inhibitors	1. Inhibitors containing a mixed salt solution of sodium chloride, potassium chloride, and NH4Cl, respectively	Sodium chloride is more effective	By adjusting the thermodynamic conditions of NGH, the gas production is inhibited	Strong stability; versatility; not limited by other mechanical conditions
	2. Different concentrations (1.5wt % and 3.0wt %) of sodium chloride solution on hydrate	Only under the condition of pure water is the sodium chloride solution effective		
	3. The hydrate formation rate under the concentration of ethylene glycol was 30.0 wt %	When the concentration of ethylene glycol is 30wt %, the hydrate inhibition effect is better		

Table 3. *Cont.*

Inhibitor Type	Experiment Method	Experimental Result	Inhibitor Mechanism	Inhibitor Effect
Mixed inhibitors	1. A mixed solution of sodium chloride and ethylene glycol	The result of mixed inhibitors is more effective than that of a single sodium chloride inhibitor	By using a mixed inhibitor of sodium chloride and ethylene glycol, the hydrate formation rate and wellbore blockage are slowed down	Strong inhibition effect; high economic type and adaptation to various environments
	2. The hydrate inhibition effect of MeOH and sodium chloride at different ratios	The effect of mixed inhibitors is better than the sum of single inhibitors		
	3. The common inhibitory effect of ethylene glycol and kinetic inhibitor	The mixing suppression effect is stronger		
	4. Hydrate inhibitor HAY	The mixed inhibitor of Tetrahydrofuran and N-butenyl pyrrolidone is more effective than a single inhibitor		
Kinetic hydrate inhibitor	1. The performance of different inhibitors was compared in a slow constant cooling experiment	Polymers containing dibutylamine groups are more potent inhibitors than polymers containing dimethylamine or diethylamine groups	It is economical and effective, environmentally friendly, and has little impact on the production process	The suppression effect is affected by many factors such as temperature and pressure, and there are differences between the laboratory evaluation method and the actual scene
	2. The effect of modified starch in inhibiting gas hydrates was studied	The results showed that, when the dosage of modified starch was 0.04wt %, the inhibition effect was the best		
	3. To study and compare the inhibitory effects of four kinds of amino acids, including aspartic acid, threonine, arginine, and valine, on NGH	Aspartic acid and threonine showed inhibitory effects, while the latter two amino acids played a promoting role		
	4. The effects of diethylene glycol butyl ether as a synergistic agent on four kinetic inhibitors, PVP, PVCap, VP/VC, and VC-713, were studied	Diethylene glycol butyl ether can significantly improve the efficiency of inhibitors		

3.6. Use a Low-Temperature Drilling Fluid to Keep Wellbore Stable

It is widely accepted that the temperature difference (ΔT) between drilling fluid and the stratum will trigger wellbore instability. Li et al. [99] have simulated the relationship between the status of NGH saturation, the variation of stratum mechanical properties, and wellbore stability (Figure 9). As the temperature of the drilling fluid gradually increases, the extent of the dissociated region of NGH around the borehole will expand. The results of the study show that, when the temperature difference ΔT reaches 7 K, the deformation degree of the formation will reach the maximum value. Based on this, the use of low-temperature drilling fluid in the NGH extraction process helps to reduce the scope of NGH dissociation and improve the stability of the wellbore during the drilling process.

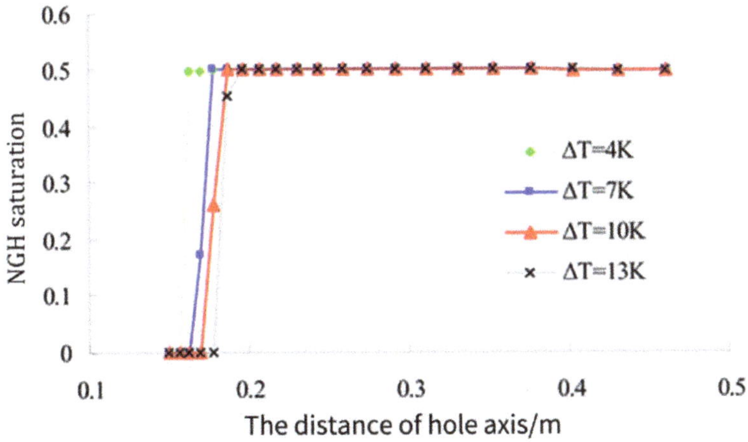

Figure 9. The distribution of hydrates saturation with ΔT ([99], reproduced with permission from Li, L et al.; [*Procedia Engineering*]; published by Science Direct [2011]).

4. Policy Suggestions

In addition to the above NGH drilling technology, the following policy suggestions are proposed to reduce the potential geological hazards.

4.1. Founding a Risk Identification Mechanism Through Marine Science and Technology Research

Marine science and technology are decisive factors in sustainable and low-risk submarine mineral resource development. Research pays much attention to advancement in two aspects: deep-sea drilling technology and deep-sea intelligent mining vehicles. Taking the latter as an example, laser radars are installed in front of intelligent mining vehicles to scan the seafloor surface, and sense rocks and other objects, and cameras are used to capture images for route planning and obstacle avoidance. At the same time, real-time positioning and map construction technology is applied to send its location and mapping information. Detailed data in the mining area can be provided for the deep-sea drilling equipment.

A risk intelligence analysis and decision-making system can effectively reduce the NGH mining risks (Figure 10). Firstly, the system supports weather forecasting and intelligent decision-making, as most of the international drilling accidents are caused by bad weather. Secondly, the system is a database of marine geology. Digital sensing technology is used to obtain raw data from NGH-bearing layers. Most importantly, the system is also a risk identification platform. It relies on big data to simulate the changes in the target area, predicts various types of geological hazards, judges the probability of potential risks, and takes appropriate measures.

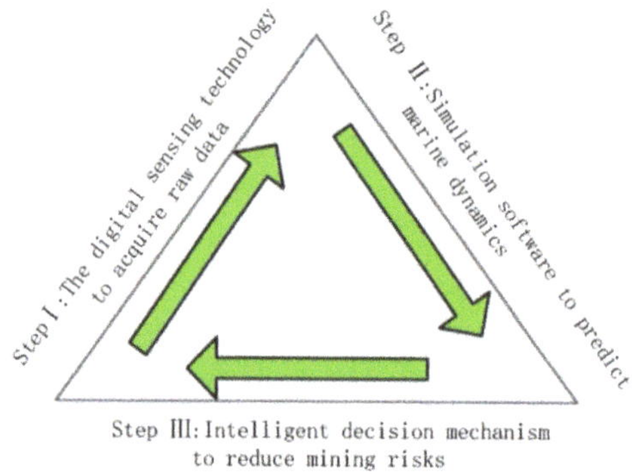

Figure 10. Intelligent risk identification and decision-making mechanism.

4.2. Establishing a Risk Warning Mechanism Through International Cooperation

Due to the breakthrough in submarine NGH mining technology, the cross-national feature of disasters, and the sensitivity of the mining sites, it is urgently required to establish an international cooperation platform. The type, scale, location, and duration of geological disasters are different. All countries should share their experiences in geological risk precautions.

Currently, international organizations believe that the heavy casualties and economic losses in the Southeast Asia tsunami in 2004 were closely related to the lack of an early warning system. An innovative early warning system has been proposed. The system includes a marine communication system, a marine early warning command office, a monitoring system for drilling tools, and appropriate back-up means, including a marine communication system, a marine early warning command office, and a monitoring system for drilling tools.

The designed mechanism (Figure 11) is centered on the marine early warning command office, connected with the submarine drilling platform by direct-line phones. Deep-sea early warning floats, seismic-with-drilling (SWD) tools, and other instruments are installed under the drilling tools to obtain seafloor stratum parameters, geological conditions, and other real-time data. According to the data, geological hazards will be accurately predicted, and corresponding measures will be taken. In the case of minor submarine landslides, an alternate foaming cement slurry program will be activated. When the red warning occurs, an emergency evacuation program should be adopted.

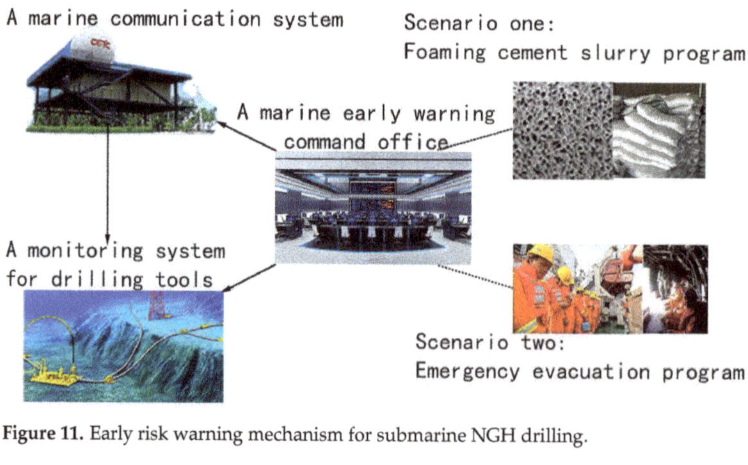

Figure 11. Early risk warning mechanism for submarine NGH drilling.

4.3. Improving Safety Regulations Through Legal Supervision

The United Nations Convention on the Law of the Sea (UNCLOS) has been widely recognized as the constitutional law of the sea since 1982. The International Seabed Authority (ISA) was established in 1994 to regulate mining and related activities in the international seafloors beyond national jurisdiction, an area that includes most of the world's oceans. However, there is still a blank in the Convention regarding NGH mining. In the future, it is necessary to clarify the mining scope and related responsibilities through international legislation. After registering the ownership of NGH in different regions, the authority of each country should be stated. It is strictly forbidden for any country to mine submarine NGH without taking any risk precautions.

Strict implementation of production safety is also an important safeguard against geological risks. In July 1988, the British Petroleum platform "Parpo Alpha" exploded because of natural gas leakage during the drilling process. The accident would not have happened if the rig had been overhauled before and during production.

5. Conclusions

NGH will be an important replacement energy source in the future due to its abundant reserves and clean nature. However, NGH may cause various natural geological hazards and secondary geological accidents during the extraction process. Some preliminary technical precautions and policy suggestions have been proposed to reduce the potential risks from the viewpoint of Earth system science systems (Table 4).

Table 4. Geological hazards in NGH mining with precautions and policy suggestions.

Geological Hazards		Technical Precautions	Policy Suggestions
Natural geological hazards	Stratum subsidence	Optimize CO_2-EGR method to alleviate stratum subsidence	1. Founding a risk identification mechanism through marine science and technology research; 2. Establishing a risk warning mechanism through international cooperation; 3. Improving safety regulations through legal supervision.
	Seafloor landslide	Fill foamed cement to control seafloor landslides	
	Greenhouse effect	Improve drilling technology to mitigate greenhouse effect	
Secondary geological accidents	Sand piping	Install new solid–liquid separator to reduce sand piping	
	Well blowout	Adopt new inhibitor to lower well blockage	
	Wellbore instability	Use a low-temperature drilling fluid to keep wellbore stable	

(1) There are two types of potential geological risks in commercial mining: natural geological hazards, including stratum subsidence, seafloor landslides, and the greenhouse effect; and secondary geological accidents, including sand piping, well blowout, and wellbore instability.

(2) Technical precautions include optimizing the CO_2-EGR method to alleviate stratum subsidence; filling with foamed cement to control seafloor landslides; improving drilling technology to mitigate the greenhouse effect; installing a new solid–liquid separator to reduce sand piping; adopting a new inhibitor to lower well blockage; and using a low-temperature drilling fluid to keep the wellbore stable.

(3) Policy suggestions include founding a risk identification mechanism through marine science and technology research; establishing a risk warning mechanism through international cooperation; and improving safety regulations through legal supervision.

Author Contributions: Conceptualization, Z.Y. and W.H.; methodology, Q.Y.; formal analysis, Z.Y., W.H., and Q.Y.; investigation, Z.Y., W.H., and Q.Y.; resources, Z.Y.; data curation, Z.Y. and W.H.; writing—original draft preparation, Z.Y., W.H., and Q.Y.; writing—review and editing, Z.Y. and W.H.;

visualization, Q.Y.; supervision, Z.Y.; project administration, W.H. and Q.Y.; funding acquisition, Q.Y. All authors have read and agreed to the published version of the manuscript.

Funding: This research was funded by Natural Science Foundation of China (No. 41962012) and Geology and Mineral Resources Survey Project (No. DD20243224).

Data Availability Statement: No new data were created or analyzed in this study. Data sharing is not applicable to this article.

Conflicts of Interest: The authors declare no conflicts of interest.

References

1. Dong, X.C. Distribution of global proven oil reserves: Historical evolution and data analysis. *China Pet. Enterp.* **2024**, *41*, 13–15.
2. Dai, R.; Huang, H.L.; Luo, M.; Li, W.T.; Wu, Y.H.; Li, X.; Zhou, S.Q. Applicability analysis of thermodynamic model of gas hydrate phase equilibrium. *Energy Chem. Ind.* **2025**, *46*, 42–50.
3. Ma, X.; Sun, Y.; Guo, W.; Jia, R.; Li, B. Numerical simulation of horizontal well hydraulic fracturing technology for gas production from hydrate reservoir. *Appl. Ocean Res.* **2021**, *112*, 102674. [CrossRef]
4. Wang, Q.; Wang, Z.; Li, P.; Song, Y.; Wang, D. Numerical modeling of coupled behavior of gas production and mechanical deformation of gas hydrate reservoir in Shenhu area, South China Sea: Enlightenments for field monitoring and model verification. *Energy* **2022**, *254*, 124406. [CrossRef]
5. Yang, Y.; Li, Q.; Xu, Y.; Huang, J. Dynamics study of self-pulling & self-rotating jet drill bit in natural gas hydrate reservoirs radial horizontal well drilling. *Geoenergy Sci. Eng.* **2025**, *244*, 213490.
6. Faramawy, S.; Zaki, T.; Sakr, A.E. Natural gas origin, composition, and processing: A review. *J. Nat. Gas Sci. Eng.* **2016**, *34*, 34–54. [CrossRef]
7. Ansari, U.; Cheng, Y.; Li, Q.; Mawaipopo, G.G.; Wei, J. Simulating the effect of subsurface stresses and transient pore pressure on wellbore stability in subsea horizontal wells. *Energy Sources Part A Recovery Util. Environ. Eff.* **2019**, *41*, 2028–2038. [CrossRef]
8. Wang, F.; Shen, K.; Zhang, Z.; Zhang, D.; Wang, Z.; Wang, Z. Numerical simulation of natural gas hydrate development with radial horizontal wells based on thermo-hydrochemistry coupling. *Energy* **2023**, *272*, 127098. [CrossRef]
9. Wang, Z.; Lei, X.; Zhou, W.; Wang, Y.; Cao, J.; Wang, C. Numerical simulation of the damage process of rock containing cracks by impacts of steel-particle water jet. *Powder Technol.* **2023**, *422*, 118465. [CrossRef]
10. Boswell, R.; Collett, T. Current perspectives on gas hydrate resources. *Energy Environ. Sci.* **2011**, *4*, 1206–1215. [CrossRef]
11. Liu, H.; Liu, H.; Zhang, Q.; Fan, M.; Yin, B.; Wang, X.; Sun, X.; Wang, Z. Effect of liquid viscosity on the gas-liquid two phase countercurrent flow in the wellbore of bullheading killing. *Geoenergy Sci. Eng.* **2023**, *221*, 111274. [CrossRef]
12. Onwukwe, S.I.; Duru, U.I. Prospect of harnessing associated gas through natural gas hydrate (NGH) technology in Nigeria. *J. Pet. Gas Eng.* **2015**, *6*, 38–44. [CrossRef]
13. Konno, Y.; Masuda, Y.; Hariguchi, Y.; Kurihara, M.; Ouchi, H. Key factors for depressurization-induced gas production from oceanic methane hydrates. *Energy Fuels* **2010**, *24*, 1736–1744. [CrossRef]
14. Yamamoto, K.; Kanno, T.; Wang, X.; Tamaki, M.; Fujii, T.; Wang, X.; Pimenov, V.; Shako, V. Thermal responses of a gas hydrate-bearing sediment to a depressurization operation. *RSC Adv.* **2017**, *10*, 5554. [CrossRef]
15. Wang, B.; Dong, H.; Liu, Y.; Lv, X.; Liu, Y.; Zhao, J.; Song, Y. Evaluation of thermal stimulation on gas production from depressurized methane hydrate deposits. *Appl. Energy* **2018**, *227*, 710–718. [CrossRef]
16. Chen, C.; Meng, Y.; Zhong, X.; Nie, S.; Ma, Y.; Pan, D.; Liu, K.; Li, X.; Gao, S. Research on the influence of injection-production parameters on challenging natural gas hydrate exploitation using depressurization combined with thermal injection stimulated by hydraulic fracturing. *Energy Fuels* **2021**, *35*, 15589–15606. [CrossRef]
17. Li, Q.; Wang, F.; Wang, Y.; Zhou, C.; Chen, J.; Forson, K.; Zhang, J. Effect of reservoir characteristics and chemicals on filtration property of water-based drilling fluid in unconventional reservoir and mechanism disclosure. *Environ. Sci. Pollut. Res.* **2023**, *30*, 55034–55043. [CrossRef]
18. Liu, F.; Tun, S.; Sun, Y. A quantitative analysis for submarine slope instability of the northern South China Sea due to gas hydrate dissociation. *Chin. J. Geophys.* **2010**, *53*, 946–953.
19. Li, J.; Ye, J.; Qin, X.; Qiu, H.; Wu, N.; Lu, H.; Xie, W.; Lu, J.; Peng, F.; Xu, Z.; et al. The first offshore natural gas hydrate production test in South China Sea. *China Geol.* **2018**, *1*, 5–16. [CrossRef]
20. Song, B.; Cheng, Y.; Yan, C.; Lyu, Y.; Wei, J.; Ding, J.; Li, Y. Seafloor subsidence response and submarine slope stability evaluation in response to hydrate dissociation. *J. Nat. Gas Sci. Eng.* **2019**, *65*, 197–211. [CrossRef]
21. Song, B.; Cheng, Y.; Yan, C.; Han, Z.; Ding, J.; Li, Y.; Wei, J. Influences of hydrate decomposition on submarine landslide. *Landslides* **2019**, *16*, 2127–2150. [CrossRef]

22. Yan, C.; Ren, X.; Cheng, Y.; Song, B.; Li, Y.; Tian, W. Geomechanical issues in the exploitation of natural gas hydrate. *Gondwana Res.* **2020**, *81*, 403–422. [CrossRef]
23. Ding, J.; Cheng, Y.; Deng, F.; Yan, C.; Sun, H.; Li, Q.; Song, B. Experimental study on dynamic acoustic characteristics of natural gas hydrate sediments at different depths. *Int. J. Hydrogen Energy* **2020**, *45*, 26877–26889. [CrossRef]
24. Li, Q.; Wang, F.; Wang, Y.; Bai, B.; Zhang, J.; Lili, C.; Sun, Q.; Wang, Y.; Forson, K. Adsorption behavior and mechanism analysis of siloxane thickener for CO2 fracturing fluid on shallow shale soil. *J. Mol. Liq.* **2023**, *376*, 121394. [CrossRef]
25. Liang, H.; Song, Y.; Chen, Y. Numerical simulation for laboratory-scale methane hydrate dissociation by depressurization. *Energy Convers. Manag.* **2010**, *51*, 1883–1890. [CrossRef]
26. Zhong, X.; Pan, D.; Zhai, L.; Zhu, Y.; Zhang, H.; Zhang, Y.; Wang, Y.; Li, X.; Chen, C. Evaluation of the gas production enhancement effect of hydraulic fracturing on combining depressurization with thermal stimulation from challenging ocean hydrate reservoirs. *J. Nat. Gas Sci. Eng.* **2020**, *83*, 103621. [CrossRef]
27. Wan, Y.; Wu, N.; Hu, G.; Xin, X.; Jin, G.; Liu, C.; Chen, Q. Reservoir stability in the process of natural gas hydrate production by depressurization in the Shenhu area of the South China Sea. *Nat. Gas Ind. B* **2018**, *5*, 631–643. [CrossRef]
28. Jin, G.; Lei, H.; Xu, T.; Liu, L.; Xin, X.; Zhai, H.; Liu, C. Seafloor subsidence induced by gas recovery from a hydratebearing sediment using multiple well system. *Mar. Pet. Geol.* **2019**, *107*, 438–450. [CrossRef]
29. Huang, L.; Xu, C.; Xu, J.; Zhao, Y. Hydrate dissociation evaluation and stratum subsidence response induced by depressurization in hydrate-bearing permafrost. *Energy Fuels* **2022**, *36*, 11077–11088. [CrossRef]
30. Jiang, Y.; Zhang, R.; Ye, R.; Zhou, K.; Gong, B.; Golsanami, N. Mechanical properties of nodular natural gas hydratebearing sediment. *Adv. Geo-Energy Res.* **2024**, *11*, 41–53. [CrossRef]
31. He, Y.; Song, B.; Li, Q. Coupling submarine slope stability and wellbore stability analysis with natural gas hydrate drilling and production in submarine slope strata in the South China Sea. *J. Mar. Sci. Eng.* **2023**, *11*, 2069. [CrossRef]
32. Zhang, Z.; Zhao, Y.J.; Cai, N.; Liu, H.X.; Ma, C.H.; Liang, J.W. The effect of decomposition of natural gas hydrate on deep-water drilling. *Pet. Sci. Technol.* **2024**, *42*, 1701–1721. [CrossRef]
33. Luo, T.; Li, Y.; Madhusudhan, B.N.; Sun, X.; Song, Y. Deformation behaviors of hydrate-bearing silty sediment induced by depressurization and thermal recovery. *Appl. Energy* **2020**, *276*, 115468. [CrossRef]
34. Wang, Y.; Feng, J.; Li, X.; Zhang, Y.; Han, H. Methane hydrate decomposition and sediment deformation in unconfined sediment with different types of concentrated hydrate accumulations by innovative experimental system. *Appl. Energy* **2018**, *226*, 916–923. [CrossRef]
35. Zhang, X.; Xia, F.; Xu, C.; Han, Y. Stability analysis of near-wellbore reservoirs considering the damage of hydratebearing sediments. *J. Mar. Sci. Eng.* **2019**, *7*, 102. [CrossRef]
36. Zhou, M.; Soga, K.; Yamamoto, K.; Huang, H. Geomechanical responses during depressurization of hydrate-bearing sediment formation over a long methane gas production period. *Geomech. Energy Environ.* **2020**, *23*, 100111. [CrossRef]
37. Xue, K.; Liu, Y.; Yu, T.; Yang, L.; Zhao, J.; Song, Y. Numerical simulation of gas hydrate production in Shenhu area using depressurization: The effect of reservoir permeability heterogeneity. *Energy* **2023**, *271*, 126948. [CrossRef]
38. Ye, J.; Qin, X.; Xie, W.; Lu, H.; Ma, B.; Qiu, H.; Liang, J.; Lu, J.; Kuang, Z.; Lu, C.; et al. The second natural gas hydrate production test in the South China Sea. *China Geol.* **2020**, *3*, 197–209. [CrossRef]
39. Yu, T.; Guan, G.; Wang, D.; Song, Y.; Abudula, A. Numerical evaluation on the effect of horizontal-well systems on the long-term gas hydrate production behavior at the second Shenhu test site. *J. Nat. Gas Sci. Eng.* **2021**, *95*, 104200. [CrossRef]
40. Xiao, C.W.; Li, X.S.; Li, G.; Yu, Y.; Yu, J.X.; Lv, Q.N. Numerical analysis of production behaviors and permeability characteristics on the second gas hydrate production test in the South China Sea. *Energy Fuels* **2022**, *36*, 10960–10974. [CrossRef]
41. Zhu, Y.; Wang, P.; Pang, S.; Zhang, S.; Xiao, R. A review of the resource and test production of natural gas hydrates in China. *Energy Fuels* **2021**, *35*, 9137–9150. [CrossRef]
42. Sun, J.; Ning, F.; Zhang, L.; Liu, T.; Peng, L.; Liu, Z.; Li, C.; Jiang, G. Numerical simulation on gas production from hydrate reservoir at the 1st offshore test site in the eastern Nankai Trough. *J. Nat. Gas Sci. Eng.* **2016**, *30*, 64–76. [CrossRef]
43. Chen, L.; Feng, Y.; Kogawa, T.; Okajima, J.; Komiya, A.; Maruyama, S. Construction and simulation of reservoir scale layered model for production and utilization of methane hydrate: The case of Nankai Trough Japan. *Energy* **2018**, *143*, 128–140. [CrossRef]
44. Yu, T.; Guan, G.; Wang, D.; Song, Y.; Abudula, A. Numerical investigation on the long-term gas production behavior at the 2017 Shenhu methane hydrate production site. *Appl. Energy* **2021**, *285*, 116466. [CrossRef]
45. Yin, F.; Gao, Y.; Chen, Y.; Sun, B.; Li, S.; Zhao, D. Numerical investigation on the long-term production behavior of horizontal well at the gas hydrate production site in South China Sea. *Appl. Energy* **2022**, *311*, 118603. [CrossRef]
46. Ma, Y.Y. *Study on Sealing Transformation and Injection Inhibitor Mining of Permeable Cap Rock in Marine Hydrate Reservoir*; Jilin University: Jilin, China, 2023.
47. Ma, S.; Li, H.; Jia, H. Influence of Natural Gas Hydrate's Property Weakening from Dissociation on Subsidence of Seabed Sediments: A Simulation Study. *Adv. Civ. Eng.* **2022**, *15*, 6159470. [CrossRef]

48. Li, S.X.; Ding, S.Y.; Wu, D.D.; Wang, X.P.; Hao, Y.M.; Li, Q.P.; Pang, W.X. Analysis of Stratum Subsidence Induced by Depressurization at an Offshore Hydrate-Bearing Sediment. *Energy Fuels* **2021**, *35*, 1381–1388. [CrossRef]
49. Zhou, S.W.; Li, Q.P.; Zhu, J.L.; Pang, W.X.; He, Y.F. Challenges and reflections on the development of gas hydrate in the South China Sea. *Nat. Gas Ind.* **2023**, *43*, 152–163.
50. Huang, Y.; Cheng, J.; Wang, M.; Wang, S.; Yan, W. Gas hydrate dissociation events during LGM and their potential trigger of submarine landslides: Foraminifera and geochemical records from two cores in the northern South China sea. *Front. Earth Sci* **2022**, *10*, 876913. [CrossRef]
51. Ma, X.; Jiang, Y.; Yan, P.; Luan, H.; Wang, C.; Shan, Q.; Cheng, X. A review on submarine geological risks and secondary disaster issues during natural gas hydrate depressurization production. *J. Mar. Sci. Eng.* **2024**, *12*, 840. [CrossRef]
52. Fell, R. Landslide risk assessment and acceptable risk. *Can. Geotech. J.* **1994**, *31*, 261–272. [CrossRef]
53. Fell, R.; Corominas, J.; Bonnard, C.; Cascini, L.; Leroie, E.; Savage, W.Z. Guidelines for landslide susceptibility, hazard and risk zoning for land use planning. *Eng. Geol.* **2008**, *102*, 85–98. [CrossRef]
54. McAdoo, B.G.; Watts, P. Tsunami hazard from submarine landslides on the Oregon continental slope. *Mar. Geol.* **2004**, *203* 235–245. [CrossRef]
55. Masson, D.G.; Harbitz, C.B.; Wynn, R.B.; Pedersen, G.; Lovholt, F. Submarine landslides: Processes, triggers and hazard prediction. *Philos. Trans. R. Soc. A Mathematical. Phys. Eng. Sci.* **2006**, *364*, 2009–2039. [CrossRef] [PubMed]
56. Kvenvolden, K.A. Potential effects of gas hydrate on human welfare. *Proc. Natl. Acad. Sci. USA* **1999**, *96*, 3420–3426. [CrossRef] [PubMed]
57. Pape, T.; Bahr, A.; Klapp, S.A.; Abegg, F.; Bohrmann, G. High-intensity gas seepage causes rafting of shallow gas hydrates in the southeastern Black Sea. *Earth Planet Sci. Lett.* **2011**, *307*, 35–46. [CrossRef]
58. Zou, H.; Li, Q.X.; Chen, A.Q.; Xiao, B.; Jiang, X.W.; Huang, C.C.; Hu, C.H.; Li, D. The Formation and Disappearance of the Neoproterozoic "Snowball Earth" and the Records from the South China Plate. *J. Sediment.* **2022**, *40*, 1043–1058.
59. Zhang, Z.G.; Wang, Y.; Gao, L.F.; Zhang, Y.; Liu, C.S. Marine gas hydrates: Future energy or environmental killer? *Energy Procedia* **2012**, *16*, 933–938. [CrossRef]
60. Sun, M.Q.; Wang, Y.Q.; Fu, P. Relationship between natural gas hydrate and global climate change. *Environ. Sci. Manag.* **2012**, *37*, 80–84.
61. Kretschmer, K.; Biastoch, A.; Rüpke, L.; Burwicz, E. Modeling the fate of methane hydrates under global warming. *Glob. Biogeochem. Cycles* **2015**, *29*, 610–625. [CrossRef]
62. Yan, C.; Li, Y.; Cheng, Y.; Wang, W.; Song, B.; Deng, F.; Feng, Y. Sand production evaluation during gas production from natural gas hydrates. *J. Nat. Gas Sci. Eng.* **2018**, *57*, 77–88. [CrossRef]
63. Zhu, N. Analysis of environmental impact factors and monitoring status of test mining of combustible ice. *Mech. Electr. Eng. Technol.* **2019**, *48*, 134–135+214.
64. Chen, K.; Wei, X.; Li, H.; Lin, H.; Khan, F. Operational risk analysis of blowout scenario in offshore drilling operation. *Process Saf. Environ. Prot.* **2021**, *149*, 422–431. [CrossRef]
65. Johansen, Ø. Development and verification of deep-water blowout models. *Mar. Pollut. Bull.* **2003**, *47*, 360–368. [CrossRef]
66. Yapa, P.D.; Chen, F. Behavior of oil and gas from deepwater blowouts. *J. Hydraul. Eng.* **2004**, *130*, 540–553. [CrossRef]
67. Sun, B.W.; Song, R. Application of a seven-component multiphase flow model to deepwater well control. *Acta Petrolei Sinica* **2011**, *32*, 1042.
68. Chen, Y.J.; Deng, C.G.; Ma, T.S. Reliability theory evaluation method for wellbore instability risk. *Nat. Gas Ind.* **2019**, *39*, 97–104.
69. Zhang, Z.W. Quantitative Evaluation of Wellbore Instability Risk in High Temperature and High Pressure Wells. Master's Thesis, Xi'an University of Petroleum, Xi'an, China, 2023.
70. Huang, Y.; Yang, J.; Shi, S.S.; Luo, M.; Yin, Q.S.; Xu, D.S. Application of managed pressure drilling technology in offshore ultra-high temperature and high pressure wells. *Pet. Drill. Process* **2018**, *40*, 699–705.
71. Chen, G.; Ewy, R.T. Investigation of the undrained loading effect and chemical effect on shale stability. In Proceedings of the SPE/ISRM Rock Mechanics Conference 2002, Irving, TX, USA, 20–23 October 2022; SPE: Calgary, AB, Canada, 2022; p. SPE-78164.
72. Bradley, W.B. Failure of inclined boreholes. *J. Energy Resour. Technol.* **1979**, *101*, 232–239. [CrossRef]
73. Chukwuemeka, A.O.; Amede, G.; Alfazazi, U. A Review of wellbore instability during well construction: Types, causes, prevention and control. *Pet. Coal* **2017**, *7*, 590–610.
74. Medhi, N.; Borthakur, P.P. An Extensive Review on Gas Hydrates: Recent Patents, Properties, Formation, Detection, Production, Importance, and Challenges. *Recent Pat. Eng.* **2025**, *19*, E080524229746. [CrossRef]
75. Yang, Y.T. *Research on Accidental Catastrophe Evolution and Early Warning Method for NGH Test Mining Operation*; China University of Petroleum (East China): Dongying, China, 2022.
76. Chen, J. Methods and risk control of NGH mining in the sea. *Sci. Technol. Wind.* **2020**, *33*, 109–110.
77. Zhao, B. *Analysis of NGH Mining Activities in the Sea Area and Potential Disaster Prevention and Control*; China University of Mining and Technology: Xuzhou, China, 2019.

78. Xin, X.; Wang, H.; Luo, J.; Yu, H.; Yuan, Y.; Xia, Y.; Chen, Q. Simulation–optimization coupling model for the depressurization production of marine natural gas hydrate in horizontal wells based on machine learning method. *Nat. Gas Ind.* **2020**, *40*, 149–158.
79. Lin, T.K.; Hsieh, B.Z. Prevention of seabed subsidence of class-1 gas hydrate deposits via CO_2-EGR: A numerical study with coupled geomechanics-hydrate reaction-multiphase fluid flow model. *Energies* **2020**, *13*, 1579. [CrossRef]
80. Wang, Z.G.; Li, X.Y.; Zhang, Y.B.; Yin, H.; Hu, C.; Liang, J.Q.; Huang, W. Analysis of the stimulation methods for marine non-diagenetic natural gas hydrate reservoirs. *Drill. Eng.* **2021**, *48*, 32–38.
81. Yang, Y.; He, Y.; Zheng, Q. An analysis of the key safety technologies for natural gas hydrate exploitation. *Adv. Geo-Energy Res.* **2017**, *1*, 100–104. [CrossRef]
82. Tian, H.; Yu, Z.; Xu, T.; Xiao, T.; Shang, S. Evaluating the recovery potential of CH4 by injecting CO_2 mixture into marine hydrate-bearing reservoirs with a new multi-gas hydrate simulator. *J. Clean. Prod.* **2022**, *361*, 132270. [CrossRef]
83. Nie, Q.; Zhang, S.; Huang, Y.; Yi, X.; Wu, J. Numerical and Experimental Investigation on Safety of Downhole Solid–Liquid Separator for Natural Gas Hydrate Exploitation. *Energies* **2022**, *15*, 5649. [CrossRef]
84. Uchida, S.; Klar, A.; Yamamoto, K. Sand production model in gas hydrate-bearing sediments. *Int. J. Rock Mech. Min. Sci.* **2016**, *86*, 303–316. [CrossRef]
85. Cha, M.J.; Hu, Y.; Sum, A.K. Methane hydrate phase equilibria for systems containing NaCl, KCl, and NH_4Cl. *Fluid Phase Equilib.* **2016**, *413*, 2–9. [CrossRef]
86. Chong, Z.R.; Chan, A.H.M.; Babu, P.; Yang, M.J.; Linga, P. Effect of NaCl on methane hydrate formation and dissociation in porous media. *J. Nat. Gas Sci. Eng.* **2015**, *27*, 178–189. [CrossRef]
87. Cha, M.; Shin, K.; Kim, J.; Chang, D.; Seo, Y.; Lee, H.; Kang, S.P. Thermodynamic and kinetic hydrate inhibition performance of aqueous ethylene glycol solutions for natural gas. *Chem. Eng. Sci.* **2013**, *99*, 184–190. [CrossRef]
88. Kwak, G.H.; Lee, K.H.; Hong, S.Y.; Seo, S.D.; Lee, J.D.; Lee, B.R.; Sum, A.K. Phase Behavior and Raman Spectroscopic Analysis for CH4 and CH4/C3H8 Hydrates Formed from NaCl Brine and Monoethylene Glycol Mixtures. *J. Chem. Eng. Data* **2018**, *63*, 2179–2184. [CrossRef]
89. Yang, L.M.; Gao, F.M.; Gao, X.; Yan, Z.J.; Xu, D.D. Development and performance evaluation of a new hydrate inhibitor HAY. *Liaoning Chem.* **2025**, *54*, 248–251.
90. Qin, H.B. *Research on Hydrate Kinetic Inhibition Mechanism and Development of Efficient Hydrate Kinetic Inhibitors*; China University of Petroleum: Beijing, China, 2016.
91. Jager, M.; Peters, C.; Sloan, E. Experimental determination of methane hydrate stability in methanol and electrolyte solutions. *Fluid Phase Equilib.* **2002**, *193*, 17–28. [CrossRef]
92. Jinhao, S.; Zhi, W.; Xuanji, L.; Xuanwei, Z.; Yumo, Z.; Shangfei, S.; Bohui, S.; Jing, G.; Xia, L. Paired KHI-MEG for synergistic inhibition of methane hydrate reformation. Chem. *Ind. Eng. Prog.* **2022**, *41*, 5373–5380.
93. Kelland, M.A.; Pomicpic, J.; Ghosh, R. Maleic and Methacrylic Homopolymers with Pendant Dibutylamine or Dibutylamine Oxide Groups as Kinetic Hydrate Inhibitors. *ACS Omega* **2022**, *7*, 42505–42514. [CrossRef]
94. Olabisi, O.T.; John, C.U.; Udim, M.C. Experimental investigation of modified starch from white corn as a kinetic inhibitor of gas hydrate. *Pet. Coal* **2019**, *9*, 1487–1493.
95. Longinos, S.N.; Longinou, D.D. Examination of different amino acids as methane-propane gas hydrate kinetic inhibitors in upstream industry. In Proceedings of the 13th Panhellenic Scientific Conference in Chemical Engineering (PESXM), Patra, Greece, 2–4 June 2022.
96. Li, B.B.; Sun, Q.; Gao, W.W. Experimental study on the effect of diethylene glycol butyl ether on the performance of kinetic hydrate inhibitors. *Sci. Technol. Eng.* **2014**, *14*, 141–144.
97. Luan, H.; Liu, M.; Shan, Q.; Jiang, Y.; Yan, P.; Du, X. Experimental Study on the Effect of Mixed Thermodynamic Inhibitors with Different Concentrations on Natural Gas Hydrate Synthesis. *Energies* **2024**, *17*, 2078. [CrossRef]
98. Bu, Y.; Du, W.; Du, J.; Zhou, A.; Lu, C.; Liu, H.; Guo, S. The potential utilization of lecithin as natural gas hydrate decomposition inhibitor in oil well cement at low temperatures. *Constr. Build. Mater.* **2021**, *269*, 121274. [CrossRef]
99. Li, L.; Cheng, Y.; Zhang, Y.; Cui, Q.; Zhao, F. A fluid-solid coupling model of wellbore stability for hydrate bearing sediments. *Procedia Eng.* **2011**, *18*, 363–368. [CrossRef]

MDPI AG
Grosspeteranlage 5
4052 Basel
Switzerland
Tel.: +41 61 683 77 34

Processes Editorial Office
E-mail: processes@mdpi.com
www.mdpi.com/journal/processes

www.ingramcontent.com/pod-product-compliance
Lightning Source LLC
Chambersburg PA
CBHW062041220926
51850CB00010B/2625
* 9 7 8 3 7 2 5 8 6 1 1 7 0 *